THE HUMAN SIDE OF ANIMALS

ROYAL DIXON

ZINC READ

Published by
This edition copyright © 2024 by Zinc Read
Design and setting By Zinc Read
Email- zincread@gmail.com
ISBN-978-93-6277-203-9

CONTENTS

FOREWORD

"And in the lion or the frog--
In all the life of moor or fen--
In ass and peacock, stork and dog,
He read similitudes of men."

More and more science is being taught in a new way. More and more men are beginning to discard the lumber of the brain's workshop to get at real facts, real conclusions. Laboratories, experiments, tables, classifications are all very vital and all very necessary but sometimes their net result is only to befog and confuse. Occasionally it becomes important for us to cast aside all dogmatic restraints and approach the wonders of life from a new angle and with the untrammelled spirit of a little child.

In this book I have attempted to bring together many old and new observations which tend to show the human-like qualities of animals. The treatment is neither formal nor scholastic, in fact I do not always remain within the logical confines of the title. My sole purpose is to make the reader self-active, observative, free from hide-bound prejudice, and reborn as a participant in the wonderful experiences of life which fill the universe. I hope to lead him into a new wonderland of truth, beauty and love, a land where his heart as well as his eyes will be opened.

In attempting to understand the animals I have used a method a great deal like that of the village boy, who when questioned as to how he located the stray horse for which a reward of twenty dollars had been offered, replied, "I just thought what I would do if I were a horse and where I would go--and there I went and found him." In some such way I have tried to think why animals do certain things, I have studied them in many places and under all conditions, and those acts of theirs which, if performed by children, would come under the head of wisdom and intelligence, I have classified as such.

Life is one throughout. The love that fills a mother's heart when she sees her first-born babe, is also felt by the mother bear, only in a different way, when she sees her baby cubs playing before her humble cave dwelling. The sorrow that is felt by the human heart when a beloved one dies is experienced in only a little less degree by an African ape when his mate is shot dead by a Christian missionary. The grandmother sheep that watches her numerous little lamb grandchildren on the hillside, while their mothers are away grazing, is just as mindful of their care as any human grandparent could be. One drop of water is like the ocean; and love is love.

The trouble with science is that too often it leaves out love. If you agree that we cannot treat men like machines, why should we put animals in that class? Why should we fall into the colossal ignorance and conceit of cataloging every human-like action of animals under the word "instinct"? Man delights in thinking of himself as only a little lower than the angels. Then why should he not consider the animals as only a little lower than himself? The poet has truly said that "the beast is the mirror of man as man is the mirror of God." Man had to battle with animals for untold ages before he domesticated and made servants of them. He is just beginning to learn that they were not created solely to furnish material for sermons, nor to serve mankind, but that they also have an existence, a life of their own.

Man has long preached this doctrine that he is not an animal, but a kinsman of the gods. For this reason, he has claimed dominion over animal creation and a right to assert that dominion without restraint. This anthropocentric conceit is the same thing that causes one nation to think it should rule the world, that the sun and moon were made only for the laudable purpose of giving light unto a chosen few, and that young lambs playing on a grassy hillside, near a cool spring, are just so much mutton allowed to wander over man's domain until its flavour is improved.

It is time to remove the barriers, once believed impassable, which man's egotism has used as a screen to separate him from his lower brothers. Our physical bodies are very similar to theirs except that ours are almost always much inferior. Merely because we have a superior intellect which enables us to rule and enslave the animals, shall we deny them all intellect and all feeling? In the words of that remarkable naturalist, William J. Long, "To call a thing intelligence in one creature and reflex action in another, or to speak of the same thing as love or kindness in one and blind impulse in the other, is to be blinder ourselves than the impulse which is supposed to govern animals. Until, therefore, we have some new chemistry that will ignore atoms and the atomic law, and some new psychology that ignores animal intelligence altogether, or regards it as under a radically different law from our own, we must apply what we know of ourselves and our own motives to the smaller and weaker lives that are in some distant way akin to our own."

It is possible to explain away all the marvellous things the animals do, but after you have finished, there will still remain something over and above, which quite defies all mechanistic interpretation. An old war horse, for instance, lives over and over his battles in his dreams. He neighs and paws, just as he did in real battle; and cavalrymen tell us that they can sometimes understand from their horses when they are dreaming just what command they are trying to obey. This is only one of the myriads of animal phenomena which man does not understand. If you doubt it, try to explain the striking phenomena of luminescence, hybridization, of eels surviving desiccation for fourteen years, post-matrimonial cannibalism, Nature's vast chain of unities, the suicide of lemmings, why water animals cannot get wet, transparency of animals, why the horned toad shoots a stream of blood from his eye when angry. If you are able to explain these things to humanity, you will be classed second only to Solomon. Yet the average scientist explains them away, with the ignorance and loquaciousness of a fisher hag.

By a thorough application of psychological principles, it is possible to show that man himself is merely a machine to be explained in terms of neurones and nervous impulses, heredity and environment and reactions to outside stimuli. But who is there who does not believe that there is more to a man than that?

Animals have demonstrated long ago that they not only have as many talents as human beings, but that under the influence of the same environment, they form the same kinds of combinations to defend themselves against enemies; to shelter themselves against heat and cold; to build homes; to lay up a supply of food for the hard seasons. In fact, all through the ages man has been imitating the animals in burrowing through the earth, penetrating the waters, and now, at last, flying through the air.

When a skunk bites through the brains of frogs, paralysing but not killing them, in order that he may store them away in his nursery-pantry so that his babes may have fresh food; when a mole decapitates earth-worms for the same reason and stores them near the cold surface of the ground so that the heads will not regrow, as they would under normal conditions, only a deeply prejudiced man can claim that no elements of intelligence have been employed.

There are also numerous signs, sounds and motions by which animals communicate with each other, though to man these symbols of language may not always be understandable. Dogs give barks indicating surprise, pleasure and all other emotions. Cows will bellow for days when mourning for their dead. The mother bear will bury her dead cub and silently guard its grave for weeks to prevent its being desecrated. The mother sheep will bleat most pitifully when her lamb strays away. Foxes utter expressive cries which their children know full well. The chamois, when frightened, whistle; they might be termed the policemen of the animal world. The sentinel will continue a long, drawn-out whistle, as long as he can without taking a breath. He then stops for a brief moment, looks in all directions, and begins blowing again. If the danger comes too near, he scampers away.

In their ability to take care of their wounded bodies, in their reading of the weather and in all forms of woodcraft, animals undoubtedly possess superhuman powers. Even squirrels can prophesy an unusually long and severe winter and thus make adequate preparations. Some animals act as both barometers and thermometers. It is claimed that while frogs remain yellow, only fair weather may be expected, but if their colour changes to brown, ill weather is coming.

There is no limit to the marvellous things animals do. Elephants, for example, carry leafy palms in their trunks to shade themselves from the hot sun. The ape or baboon who puts a stone in the open oyster to prevent it from closing, or lifts stones to crack nuts, or beats his fellows with sticks, or throws heavy cocoanuts from trees upon his enemies, or builds a fire in the forest, shows more than a glimmer of intelligence. In the sly fox that puts out fish heads to bait hawks, or suddenly plunges in the water and immerses himself to escape hunters, or holds a branch of a bush over his head and actually runs with it to hide himself; in the wolverine who catches deer by dropping moss, and suddenly springing upon them and clawing their eyes out; in the bear, who, as told in the account of Cook's third voyage, "rolls down pieces of rock to crush stags; in the rat when he leads his blind brother with a stick" is actual reasoning. Indeed, there is nothing which man makes with all his ingenious use of tools and instruments, of which some suggestion may not be seen in animal creation.

Great thinkers of all ages are not wanting who believe that animals have a portion of that same reason which is the pride of man. Montaigne admitted that they had both thought and reason, and Pope believed that even a cat may consider a man made for his service. Humboldt, Helvitius, Darwin and Smellie claimed that animals act as a definite result of actual reasoning. Lord Brougham pertinently observes, "I know not why so much unwillingness should be shown by some excellent philosophers to allow intelligent faculties and a share of reason to the lower animals, as if our own superiority was not quite sufficiently established to leave all jealousy out of view by the immeasurably higher place which we occupy in the scale of being."

From the facts enumerated in this book I find that animals are possessed of love, hate, joy, grief, courage, revenge, pain, pleasure, want and satisfaction--that all things that go to make up man's life are also found in them. In the attempt to establish this thesis I have been led mentally and physically into some of Nature's most fascinating highways and hedges, where I have had many occasions to wonder and adore. I will be happy if I have at least added something to the depth of love and appreciation with which most men look upon the animal world.

ROYAL DIXON.

New York, April, 1918.

I

ANIMALS THAT PRACTISE CAMOUFLAGE

"She was a gordian shape of dazzling line,
Vermilion-spotted, golden, green and blue;
Striped like a zebra, freckled like a pard,
Eyed like a peacock, and all crimson barr'd,
And full of silver moons, that, as she breathed,
Dissolved, or brighter shone, or interwreathed
Their lustres with the glorious tapestries...."

--KEATS (*on Lamia, the snake*).

The art of concealment or camouflage is one of the newest and most highly developed techniques of modern warfare. But the animals have been masters of it for ages. The lives of most of them are passed in constant conflict. Those which have enemies from which they cannot escape by rapidity of motion must be able to hide or disguise themselves. Those which hunt for a living must be able to approach their prey without unnecessary noise or attention to themselves. It is very remarkable how Nature helps the wild creatures to disguise themselves by colouring them with various shades and tints best calculated to enable them to escape enemies or to entrap prey.

The animals of each locality are usually coloured according to their habitat, but good reasons make some exceptions advisable. Many of the most striking examples of this protective resemblance among animals are the result of their very intimate association with the surrounding flora and natural scenery. There is no part of a tree, including flowers, fruits, bark and roots, that is not in some way copied and imitated by these clever creatures. Often this imitation is astonishing in its faithfulness of detail. Bunches of cocoanuts are portrayed by sleeping monkeys, while even the leaves are copied by certain tree-toads, and many flowers are represented by monkeys and lizards. The winding roots of huge trees are copied by snakes that twist themselves together at the foot of the tree.

In the art of camouflage--an art which affects the form, colour, and attitude of animals--Nature has worked along two different roads. One is easy and direct, the other circuitous and difficult. The easy way is that of protective resemblance pure and simple, where the animal's colour, form, or attitude becomes like that of its habitat. In which case the animal becomes one with its environment and thus is enabled to go about unnoticed by its enemies or by its prey. The other way is that of bluff, and it includes all inoffensive animals which are capable of assuming attitudes and colours that terrify and frighten. The colours in some cases are really of warning pattern, yet they cannot be considered mimetic unless they are thought to resemble the patterns of some extinct model of which we know nothing; and since they are not found in present-day animals with unpleasant qualities, they are not, strictly speaking, warning colours.

Desert animals are in most cases desert-coloured. The lion, for example, is almost invisible when crouched among the rocks and streams of the African wastes. Antelopes are tinted like the landscape over which they roam, while the camel seems actually to blend with the desert sands. The kangaroos of Australia at a little distance seem to disappear into the soil of their respective localities, while the cat of the Pampas accurately reflects his surroundings in his fur.

The tiger is made so invisible by his wonderful colour that, when he crouches in the bright sunlight amid the tall brown grass, it is almost impossible to see him. But the zebra and the giraffe are the kings of all camouflagers! So deceptive are the large blotch-spots of the giraffe and his weird head and horns, like scrubby limbs, that his concealment is perfect. Even the cleverest natives often mistake a herd of giraffes for a clump of trees. The camouflage of zebras is equally deceptive. Drummond says that he once found himself in a forest, looking at what he thought to be a lone zebra, when to his astonishment he suddenly realised that he was facing an entire herd which were invisible until they became frightened and moved. Evidently the zebra is well aware that the black-and-white stripes of his coat take away the sense of solid body, and that the two colours blend into a light gray, and thus at close range the effect is that of rays of sunlight passing through bushes.

The arctic animals, with few exceptions, are remarkable for imitating their surroundings; their colour of white blends perfectly with the snow around them. The polar bear is the only white bear, and his home is always among the snow and ice. The arctic fox, alpine hare, and ermine change to white in winter only, because during the other seasons white would be too conspicuous. The American arctic hare is always white because he always lives among the white expanses of the Far North. Both foxes and stoats are carnivorous and feed upon ptarmigan and hares, and they must be protectively coloured that they may catch their prey.

On the other hand, Nature aids the prey by providing them with colours that enable them to escape the attention of their enemies.

The young of many of the arctic animals are covered with fluffy white hair, so that while they are too young to swim they may lie with safety upon the ground and escape the attention of polar bears; but in the antarctic regions, where there are few enemies to fear, the young seals, for instance, are exactly the colour of their parents.

The most remarkable exception of mimetic colouring among the animals of the polar regions is the sable. Throughout the long Siberian winter he retains his coat of rich brown fur. His habits, however, are such that he does not need the protection of colour, for he is so active that he can easily catch wild birds, and he can also subsist upon wild berries. The woodchuck of North America retains his coat of dark-brown fur throughout the long, cold winters. The matter of his obtaining food, however, is easy, for he lives in burrows, near streams where he can catch fish and small animals that live in or near the water.

A number of the old-school naturalists believed that when an animal's colouring assumed the snowy-white coat of its arctic surroundings, this was due to the natural tendency on the part of its hair and fur to assume the colourings and tints of their habitat. This, however, is absolutely false; and no better proof of it can be offered than the case of the arctic musk-ox, who is far more polar in his haunts than even the polar bear, and is therefore exposed to the whitening influence of the wintry regions more than the bear. Yet he never turns white, but is always brown. The only enemy of this northern-dweller is the arctic wolf, and against this enemy he is protected by powerful hoofs, thick hair, and immense horns. He does not need to conceal himself, and therefore does not simulate the colour of his surroundings.

American Museum of Natural History, New York

THE INDIANS CLAIM THAT THE MOTHER BISON FORCED HER CALF TO ROLL OFTEN IN A PUDDLE OF RED CLAY, SO THAT IT MIGHT BE INDISTINGUISHABLE AGAINST ITS RED CLAY BACKGROUND.

American Museum of Natural History, New York

THE ZEBRA IS ONE OF THE CLEVEREST OF CAMOUFLAGERS. THE BLACK-AND-WHITE STRIPES OF HIS BODY GIVE THE EFFECT OF SUNLIGHT PASSING THROUGH BUSHES.

Mimetic resemblances are worked out with great difficulty, except in such cases as the nocturnal animals, which simply become one with their surroundings. Mice, rats, moles, and bats wear overcoats that are very inconspicuous, and when suddenly approached they appear almost invisible. Some of the North American Indians claimed that buffaloes made their calves wallow in the red clay to prevent them from being seen when they were lying down in the red soil.

The kinds of protection from these mimetic resemblances are many and varied: the lion, because of his sandy-colouring, is able to conceal himself by merely crouching down upon the desert sands; the striped tiger hides among the tufts of grass and bamboos of the tropics, the stripes of his body so blending with the vertical stems as to prevent even the natives from seeing him in this position. The kudu, one of the handsomest of the antelopes, is a remarkable animal in several ways. His camouflage is so perfect that it gives him magnificent courage. With his spiral horns, white face, and striped coat tinted in pale blue, he is almost invisible when hiding in a thicket. The perfect harmony of his horns with the twisted vines and branches, and the white colourings with blue tints in the reflected sunlight conceal him entirely.

The snow-leopard, which inhabits Central Asia, is stony-grey, with large annular spots to match the rocks among which he lives. This colouration conceals him from the sheep, upon which he preys; while the spotted and blotchy pattern of the so-called clouded tiger, and the peculiarly-barred skin of the ocelot, imitate the rugged bark of trees, upon which these animals live.

One of the most unusual and skilled mimics is the Indian sloth, whose colour pattern and unique eclipsing effects seem almost incredible to those unfamiliar with the real facts. His home is in the trees, and he has a deep, orange-coloured spot on his back, which would make him very conspicuous if seen out of his home surroundings. But he is very clever, and clings to the moss-draped trees, where the effect of the orange-coloured spot is exactly like the scar on the tree, while his hair resembles the withered moss so strikingly that even naturalists are deceived.

Henry Drummond must have known the animal world rather well when he remarked that "Carlisle in his blackest visions of 'shams and humbugs' among humanity never saw anything so finished in hypocrisy as the naturalist now finds in every tropical forest. There are to be seen creatures, not singly, but in tens of thousands, whose every appearance, down to the minutest spot and wrinkle, is an affront to truth, whose every attitude is a pose for a purpose, and whose whole life is a sustained lie. Before these masterpieces of deception the most ingenious of human impositions are vulgar and transparent. Fraud is not only the great rule of life in a tropical forest, but the one condition of it."

Many of the larger cats live in trees, and most of them have spotted or oscillated skins, which aid them in hiding among foliage plants. The puma who wears a brown coat is an exception, but it must be remembered that he does not need the kind of coat his fellow friends wear. He clings so closely to the body of a tree while waiting for his prey as to be almost invisible.

This phenomenon is true throughout the animal world. Everywhere does Nature aid in escape and capture. Only those skilled in the ways of the wild fully realise how conspicuous amidst foliage, for instance, would be a uniform colouration. A parti-coloured pattern is extremely deceptive and thus protective, and for this reason one seldom sees in Nature a background of one colour; and since the large majority of animals need concealment, it is necessary for them to be clothed in patterns that vary.

These variations are especially noticeable in young animals, and furnish them with a mantle that is practically invisible to predatory enemies during the time they are left unprotected by their parents. These protective mantles often differ strikingly in pattern and colouration from those of their parents, and indicate that the young animals present the colouration and pattern of their remote forbears. It might even be said that "the skins of the fathers are thrust upon the children, even unto the third and fourth generation!" In fact, it is quite probable that they give through this varying colouration the "life-history" of their family.

In all hoofed animals--antelope, deer, horses--the protective colouration is also adapted to habitat and environment. Most deer belong to the forest, carefully avoiding the open deserts and staying near water. They live chiefly in the jungle or scrub, and are usually spotted with red and white in such a way as to be almost invisible to a casual observer; some, however, that live in the very shady places are uniformly dark so as to harmonise with their surroundings. The wild horses and asses of Central Asia are dun-coloured--corresponding exactly to their sandy habitat.

The Shakesperian conception of the human world as a stage may be paralleled in the animal world. Animals, like human beings, have all a definite role to play in the drama of life. Each is given certain equipment in form, colour, voice, demeanour, ambitions, desires, and natural habitat. Some are given much, others but little. Many have succeeded well in the art of camouflage while endeavouring to make a success in life. This success has brought the desired opportunity of mating, rearing young, bequeathing to them their special gifts and living in ease and comfort.

One of the most successful and striking cases of protective colouration in young animals is found in wild swine. Here there is longitudinal striping which marks them from head to tail in broad white bands, over a background of reddish dark brown. The tapirs have a most unique form of marking. It is similar in the young of the South American and Malayan species. Their bodies are exquisitely marked in snow-white bars. At their extremities these bars are broken up into small dots which tend to overlap each other. During the daytime these young animals seek the shade of the bushes and as the spots of sunlight fall upon the ground they appear so nearly one with their environment as to pass unnoticed by their enemies. The adults, however, vary greatly one from another in colouration. The American species is self-coloured, while the Malayan has the most unique pattern known to the animal world. The fore-quarters, the head, and the hind-legs are black, while the rest of the body from the shoulders backwards is of a dirt-white colour.

It has been observed by all students of Nature that bold and gaudy animals usually have means of defending themselves that make them very disagreeable to their enemies. They either have poisonous fangs, sharp spines, ferocious claws, or disagreeable odours. There are still others that escape destruction because of the bad company with which they are associated by their enemies.

The reptiles offer us many good examples of mimicry. Most arboreal lizards wear the colour of the leaves upon which they feed; the same is true of the whip-snakes and the tiny green tree-frogs. A striking example of successful camouflage is found in the case of a North American frog whose home is on lichen-covered rocks and walls, which he so closely imitates in colour and pattern as to pass unnoticed so long as he remains quiet. I have seen an immense frog, whose home was in a damp cave, with large green and black spots over his body precisely like the spots on the sides of his home.

Author Note: **The word "mimicry" as used here implies a particular kind of resemblance only, a resemblance in external appearance, never internal, a resemblance that deceives. It does not imply voluntary imitation. Both the words "mimicry" and "imitation" are used to imply outward likeness. The object of the outward likeness or resemblance is to cause a harmless or unprotected animal to be mistaken for the dangerous one which he oftentimes imitates; or to aid the unprotected animal in escaping unnoticed among the surroundings he may simulate.**

A splendid example of pure bluff is shown in the case of the harmless Australian lizard, known scientifically under the name of *chlamydosaurus kingii*. When he is undisturbed he seems perfectly inoffensive, but when he becomes angry, he becomes a veritable fiend-like reptile. In this condition he stands up on his hind legs, opens his gaping mouth, showing the most terrible teeth, which, by the way, have never been known to bite anything. Besides this forbidding display he further adds to his terrible appearance by raising the most extraordinary frill which is exquisitely decorated in grey, yellow, scarlet, and blue. This he uses like an umbrella, and if in this way he does not succeed in frightening away his enemy, he rushes at him, and lashes him with his saw-like tail. Even dogs are terrified at such camouflage and leave the successful bluffer alone.

In all parts of the tropics are tree-snakes that lie concealed among the boughs and shrubs. Most of them are green, and some have richly coloured bands around their bodies which look not unlike gaily coloured flowers, and which, no doubt, attract flower-seeking insects and birds. Among these may be mentioned the deadly-poisonous snakes of the genus *elaps* of South America. They are so brilliantly provided with bright red and black bands trimmed with yellow rings that it is not uncommon for a plant collector to attempt to pick them up for rare orchids!

Wherever these snakes are found, are also found a number of perfectly harmless snakes, absolutely unlike the dangerous ones in habit and life, yet coloured precisely the same. The *elaps fulvius*, for example, a deadly venomous snake of Guatemala, has a body trimmed in simple black bands on a coral-red ground, and in the same country and always with him is found a quite harmless snake, which is coloured and banded in the same identical manner. The terrible and much-feared *elaps lemnicatus* has the peculiar black bands divided into divisions of three by narrow yellow rings, thus exactly mimicking a harmless snake, the *pliocerus elapoides*, both of which live in Mexico. Presumably, the deadly variety assumes the colouring of the harmless kind in order to deceive intended victims as to his ferocity.

Surely this is sufficient evidence that colouration and pattern-design is a useful camouflage device of the great struggle for existence. And it is safe to assert that any animal that has enemies and still does not resort to protective colouration or mimicry in some form is entirely able to protect itself either by its size, strength, ferocity, or by resorting to safety in numbers. Elephants and rhinoceroses, for example, are too powerful to be molested when grown, except in the rarest cases, and are furthermore thoroughly capable of protecting their young. Hippopotamuses are protected by their immense heads, and are capable of defending their young from crocodiles even when in the water.

The bison and buffalo, which were once so powerful on the plains of North America, were protected by their gregarious habits, which terrorised their enemies--the wolves. Their nurseries were a feature of their wisdom. These were circular pens where the tall grass was tramped down by expectant mothers for the protection of their young. This natural nursery was protected from the inside by sentinels who went round and round the pen constantly guarding the young not only from the attack of wolves but also from venturing forth alone too early into the open unprotected plains. In a similar way the snow-pens of the moose of the Far North serve to protect them from the hungry hordes of wolves of which they live in constant danger. This indicates that the annihilation of the bison and buffalo was due, not to lack of wisdom, but to man's inhumanity; for, taking advantage of their nurseries, the men crouched near and concealing themselves in the grass killed not only the mothers for food but even the young in their savage sport.

The large majority of monkeys are protectively coloured with some shade of brown or grey, with specially marked faces. Entire packs of Ceylonese species will, at the slightest alarm, become invisible by crouching on a palm-tree. One of the most strikingly coloured African monkeys is jet black with a white bushy tail, and a face surrounded by a white ring, or mantle of long silky hair. He thus simulates so strikingly the hanging white lichens upon the trees that he is rarely seen by his enemies.

A book might be written upon the various ways that animals, when closely associated with other animals or human beings, imitate them. Darwin says that "two species of wolves, which had been reared by dogs, learned to bark, as does sometimes the jackall," and it is well known that certain dogs, when reared by cats, imitate their habits, even to the licking of their feet and the washing of their faces. If a mongrel dog associates with a trained dog for any period of time it is remarkable the progress he will make. For this same reason young dogs are carried on hunting trips with trained dogs that they may learn by imitation the art of hunting.

In the whole realm of Nature there is nothing more wonderful than this matter of protective colouration. Animals do not monopolise the art. It extends through the whole world of living creatures. The fact that individual animals have no voluntary control over their own colour is eloquent testimony as to the existence of mysterious life forces and racial evolutions which are still far beyond the grasp of man's understanding. To see a tiny chameleon adapt his colouring to his environment, be it red, green, or yellow, in the twinkling of an eye, is to have seen an argument for God Himself.

II

ANIMAL MUSICIANS

"Nay, what is Nature's self,
But an endless strife towards
Music, euphony, rhyme?"

--WATSON.

The great thinkers of the age believe that the world is one marvellous blending of innumerable and varied voices. This unison of sound forms the great music of the spheres, which the poets and philosophers have written so much about. Even from a purely scientific point of view, there is no denying that this music exists. Aviators tell us that when they listen from a distance to the myriads of noises and sounds that arise over a great city, these are all apparently lost in a modulated hum precisely like the vibrations of an immense tuning-fork, and appearing as but a single tone. Thus the immense noise going from our world is musically digested into one tone, and the aviator soaring above the earth hears only the one sound--the music of the spheres.

The deep appreciation that animals have for music is becoming a generally known fact among those who have studied them closely. Every one must admit that there is much truth in the old saying that "music hath charms to soothe the savage breast." Music is composed of vibrations, which act with great power upon the nervous system of men and animals alike. Each is affected according to his particular physical and mental development.

Professor Tarchanoff has made a careful study of the influence of music upon men and animals. He has demonstrated, by means of a machine which carefully registers the various activities of the hands and fingers, that when the hands are so tired and fatigued that they cannot make any marks except a straight line on the cylinder which registers the movements, music will so stimulate the nerves as to cause all fatigue to disappear. And as soon as the fingers again touch the cylinder, they begin to draw lines of various kinds and heights, thus proving that the music had rested the fingers and placed them under control. Various kinds of music were used: that of a melancholy nature had precisely the opposite effect to that of a lively, cheerful character; the nerves of the hands could either be contracted or expanded according to the nature of the music.

Like all real scientists, Professor Tarchanoff does not claim to give any positive explanation of these facts. He believes, however, that the voluntary muscles act in the same relation to the music as the heart--that is, that cheerful, happy music affects the excito-motor nerves, sets up a vibration in those nerves which produces cheer and good feeling; while sad, morbid music plays along the depressant nerves and produces sadness and depression.

In view of these facts, it is easy to see how animals, with their nervous temperaments and ready response to outside stimuli, are greatly influenced by various kinds of music. It is scientifically recognised that music tends to increase the elimination of carbonic acid and increases not only the consumption of oxygen, but even the activities of the skin. There is no doubt that good music at meal time aids the digestion.

American Museum of Natural History, New York

MONKEYS ARE THE MOST MUSICAL OF ALL ANIMALS. WHEN THEY CONGREGATE FOR "CONCERTS," AS SOME OF THE TRIBES DO, THE AIR IS FILLED WITH WEIRD STRAINS OF MONKEY-MUSIC.

CATS, UNLIKE DOGS, ARE VERY FOND OF MUSIC. AND IT HAS BEEN PROVED THAT THEIR MUSIC-SENSE CAN BE DEVELOPED TO A REMARKABLE DEGREE.

Cats have a species of unbeautiful music all their own, generally produced at late hours of the night on the house tops, garden walls, and in the alleys of our dwellings. Miss Cat's songs are far too chromatic to be appreciated by human ears; as a result her concertos and solos are rarely spoken of by human critics. However, Nature does sometimes produce a Tetrazzini, Alice Neilson, or Caruso, in the form of a cat, which really delights in harmonious combinations of sound. I know, for instance, of a cat called "Nordica" owned by Presson Miller, who apparently takes the greatest delight in hearing good vocal and instrumental music. Another well-educated musical cat belongs to a friend who plays a guitar. This cat delights in touching the strings with his dainty, soft paws, and springs with delight as the notes are produced.

The *Animal World* speaks of five musical cats, which were carried to various parts of the world and exhibited as "bell-ringers," and their owner made a fortune out of their concerts. Five bells were suspended from a hoop, which hung above the stage, and to each bell was attached a small rope. At a given signal, each cat would seize a bell and give it a pull. This was done with such perfect time and spirit that one might well believe it was the work of human musicians and not of cats.

Cows are responsive to certain kinds of music. A funeral march makes them sad, and ragtime so disturbs them that they give but little milk. The newspapers claim that Charles W. Ward, who owns a ranch near Eureka, California, says that the right kind of music will increase the production of milk, and that he uses a phonograph in the dairy barn.

A friend, who has travelled much, tells the story of a musical cow. He, in company with two other friends, was coming up a river in a small boat singing. Just as they turned a bend, they saw a small brown cow, suckling her calf, along with several other cows in a nearby pasture. The cow seemed so fascinated with the music that she plunged into the water and waded up to her head trying to reach the boat. As they rowed along, she ran up and down the bank, cutting capers in a most astonishing manner and lowing and bellowing in testimony of her delight in the music. She would leap, skip, roll on the grass, paw up the earth, like an angry bull, and chase off like a playful kitten, always with a low plaintive bellow as a final farewell. These friends often rowed up the river just to see if the musical cow was there, and she always greeted them in the usual appreciative manner.

Lions and tigers are proverbially fond of music. Professional trainers tell us that these animals, when tamed, will not do their stunts without the accompaniment of music. The story is told of a group of tigers which recently refused to perform, because the musicians, while the performance was going on, went on a strike. At once when the music ceased, the animals returned to their respective seats and no amount of encouragement would induce them to continue their performance. No amount of threats would induce them to work without music. The trainer dared not punish them too severely, yet he feared that if they were not forced to perform, they might continue to strike. But such was not the case, for on the morrow when the musicians returned they acted as never before.

Sheep, both tame and wild, are exceedingly fond of music, and the shepherds of Scotland have used it with their sheep for ages. When the shepherd plays upon his flute or bagpipe, they gather around him and listen apparently with great satisfaction; when the music ceases, they wander out to feed, and in the evening he leads them home by the single strains of his flute.

Circus horses are not only fond of music, but are partial to certain tunes, and demand that these be played while they are doing their turn. If for any reason the band changes the tune during a performance, they immediately refuse to go on with their stunts.

The original fountain of all music was based on the various voices and sounds of animals--and each musical instrument was originally devised to imitate these sounds. For all instruments--the bass drum, flute, clarinet, trombone, trumpet, violin, and even pipe organ--an animal may be mentioned that owns the fundamental tones in its voice, and which man has imitated. Castanets, for example, were imitations of the rattlesnakes; the first musical instruments of any savage tribe of men are made so as to represent the voices of the chief animals of that particular locality.

Every animal of the higher order, with the exception of a few mute dogs that belong to very hot or cold climates, is possessed of some sort of musical tone, expressive of pain or joy, and by means of which he can express certain emotions. Darwin claimed that the voice of the gibbon, while extremely loud, was very musical; and Waterhouse said that this musician sang the scale with considerable accuracy, at least sufficiently well for a trained violinist to accompany him.

Often when dogs hear music they howl, or attempt to sing. Some show a decided preference for certain kinds of music, and actually try to imitate it. Gross tells of a friend of his who had a dog with which he often gave performances. The dog would accompany his master, when he sang in falsetto, with howls that were unmistakably attempts at singing, and which readily adapted themselves to the pitch of the tone. This was a musical accomplishment of which he was very proud.

On a subject of which so little is known, there are, of course, diverse opinions. Scheitlin believed that music is actually disagreeable to a dog, but he says that it may be questioned whether or not the dog does not in some way accompany it. And Romanes, the great animal authority, thought the same thing. He had a terrier, which accompanied him when he sang, and actually succeeded in following the prolonged notes of the human voice with a certain approximation to unison. Dr. Higgins, a musician, claimed that his large mastiff could sing to the accompaniment of the organ.

Alix gives such positive examples that they are really marvellous: "Pere Pardies cites the case of two dogs that had been taught to sing, one of them taking a part with his master. Pierquin de Gembloux also speaks of a poodle that could run the scale in tune and sing very agreeably a fine composition of Mozart's *My Heart It Sings at Eve*." All the scientists in Paris, according to the same authority, went to see the dog belonging to Dr. Bennati, and hear it sing the scale, which it could do perfectly.

Monkeys and apes most nearly approximate human musicians. In central Africa these animal tribes have musical centres where they congregate regularly for "concerts." Prof. Richard S. Garner, the noted authority on apes and monkeys, believes that the time has already come for the establishment of a school for their education. He would have the courses beginning with a kindergarten and advancing through as many grades as the students required. Prof. Garner furthermore believes that we have little understanding of the gorilla, and points out that these animals have a very happy and harmonious home life, the father being highly domestic and delighting in the company of his wife and children. It is not uncommon to find five or six generations in a certain district of the jungle.

Their near kin, the chimpanzees, are equally clannish, but more musical. They come down from the branches of the trees, seating themselves on the dry leaves and assembling like an orchestra. After all are ready, they begin beating the leaves with their hands, at first very slowly, like the quiet prelude to a symphony, and gradually increasing in tempo until the grand crescendo is reached. Then, as if by the direction of an invisible leader, the music suddenly ceases. To deny that this is to them a real concert would lead us into extreme absurdities. In this connection it is interesting to note that when a baby is expected in the village, all music ceases until after its birth, when they again resume their periodic musical festivals. Hensel verifies this observation, and tells us of having seen apes come from their shelter in the early morning and congregate for a musical concert. "They repair," he says, "to the shelter of some gigantic monarch of the forest whose limbs offer facilities for walking exercises. The head of the family appropriates one of these branches and advances along it seriously, with elevated tail, while the others group themselves about him. Soon he gives forth soft single notes, as the lion likes to do when he tests the capacity of his lungs. This sound, which seems to be made by drawing the breath in and out, becomes deeper and in more rapid succession as the excitement of the singer increases. At last, when the highest pitch is reached, the intervals cease and the sound becomes a continuous roar, and at this point all the others, male and female, join in, and for fully ten seconds at a time the awful chorus sounds through the quiet forest. At the close the leader begins again with the detached sounds."

Perhaps the most remarkable evidence of animals showing a comprehensive intelligence of musical pitch is demonstrated by cavalry horses. That they thoroughly understand it is clearly demonstrated by the fact that they will obey the calls of the bugle for cavalry evolutions without a moment's hesitation and with no suggestion from outside sources. These bugle calls are produced by a combination of four notes, each of a different pitch, and it is rarer to find a horse making a mistake in the musical orders given than it is for their masters.

Rats and mice have a decided liking for music, as is attested by the fact that they appear as uninvited guests and also come as near the performer as possible. Mice, one would believe, love church music, for they often build their nests in pipe organs, thus being able to rear their children in both a musical and religious atmosphere! There is more truth than imagination in the story of the Pied Piper of Hamelin, which illustrates how they respond to the simple charms of music.

Even donkeys betray tendencies toward musical efforts, and seem to be aroused by music at least temporarily to a higher mental plane than Balaam was inclined to ascribe to his wise ass. Not all of them sing equally well, but in Arizona the donkey is known as the "desert canary." If you were to spend a few glorious days in the Hopi village of Araibi, you would hear through the still, silent night their long nasal bray or song, and you would be convinced that the term is quite appropriate. You may not exactly like the tune, but you will concede that they sing!

Society is just awakening to the joy and the significance of community art. This is everywhere indicated by the great growing group of people who come together for a common music, either as a chorus or an orchestra or both. But in this field man has not yet attained such unity of communal effort as have the frogs. In the great swamps of the world myriads of them gather from miles around, conscious of one purpose, and by a marvellous understanding and co-operation create for themselves a symphony with beauties and harmonies of its own, and such as to stand unrivalled in man's musical world. In the great chorus are voices from the lowest bass of the croaking bullfrog, squatting in the marshes, to the myriads of tiny green tree tenors, between which are millions of altos, contraltos, sopranos, coloraturas and other voices not yet in our musical vocabulary. These are accompanied by all the sounds of our orchestra and innumerable others of such delicate shades and gradations as to defy the ear of man. If we listen to one of these concerts, we will quickly recognise the tones of every familiar instrument, such as the drum, pipe, horn, trombone, oboe, piccolo, 'cello, and violin. The greatest of these musical festivals directly precedes the mating season, and is a dramatic instance of a manifestation of an inner rhythm which corresponds to an external periodicity.

Among the oldest traditions of the Eastern world are those of snake-charming by means of music. I have long been interested in this strange phenomenon of Nature, and in company with a brilliant young violinist visited a zoological park recently, and after securing permission from the head keeper, entered the snake-house. The violinist began by playing a few most sympathetic chords, first delicate and soft, then sad, then gay, slow or tremulous. Near us, coiled in his immense cage, was a large cobra--the snake which all legend claims is most easily influenced by music. Almost immediately after the music began, the cobra raised himself in a listening attitude, steadily gazed at us as though he were viewing the future, spread his immense hood, and slowly began to shake his head from side to side, as if he were trying to keep time to the music. As soon as the music would change, his attitude changed accordingly. Only after the music had ceased did he resume his normal position.

The Indians agree that under the influence of various musical instruments, especially bagpipes, snake-charmers are able to get the snakes to come out from their homes among the old rocks and walls, and when they appear they seem perfectly dazed so that they can be easily captured.

It is not well to have any kind of musical instrument played, when in a forest at night where there are dangerous snakes, lest they come to hear it. Snake-hunters always carry with them some kind of musical instrument, depending upon the kind of snakes they wish to capture. It seems that all are not equally fascinated by it. I have experimented with little effect upon a large rattler; it may have been that he was deaf. But he gave little evidence of being interested.

We need not feel humiliated, then, for our animal kinspeople with their primitive music: we were monkeys, and before them we were reptiles, birds, fishes, even worms. But that was ages ago, and we have grown up and become better musicians. Evolution has chosen us as its favourites and given us every advantage in the struggle up the ladder of life. Our musical rivals of yesterday are as chorus people compared to Metropolitan Opera stars, with us. On this earth we reign supreme, we have conquered the earth, air, and water, annihilating time and distance. What more is there for us to learn of Nature's secrets? Only an understanding of our lower brothers, the animals.

III

ANIMALS AT PLAY

"... *About them frisking, played*
All beasts of the earth, since wild, and of all chase
In wood or wilderness, forest or den;
Sporting the lion romped, and in his paw
Dandled the kid; bears, tigers, ounces, pards,
Gambled before them; the unwieldy elephant,
To make them mirth, used all his might, and wreathed
His light proboscis."

--*Paradise Lost.*

That "one touch of Nature makes the whole world kin" is shown in no clearer way than by the games and play of animals. Recreation is as common among them as it is among our own children; and they seem always to be artistic and even skilled in their play. Young goats and lambs skip, jump, run races, throw flips in the air, and gambol; calves have interesting frolics; young colts and mules have biting and kicking games; bears wrestle and tumble; puppies delight in biting and tussling; while kittens chase everything from spools of thread to their own tails.

But animal children grow up, and stop playing to a certain extent as age advances, precisely as human children do. Each settles down into a more practical condition of life. They dislike to have their games and play disturbed, and if the mother dog growls because her playful son has continuously tumbled over her while she was sleeping, or the cat-mother slaps her kitten because he plays with her tail--it is a display of the same kind of emotion that prompts a human mother to rebuke her child in the nursery for making too much noise, or for throwing toys out of the window. Animals, like ourselves, feel every sensation of joy, happiness, surprise, disappointment, love, hope, ambition, and through their youthful games an entire index of their future lives may be obtained.

This play has much to do with the physical and mental development of the animals; and it is strange indeed that so few writers have considered the subject of play in the animal world. Most of those who have noticed the subject at all, drop it with a few remarks, to the effect that it is "highly amusing," or "very funny," or "unbelievable," or "so like the play of children," without even a word of explanation of the whys and wherefores of it.

All animals have some kinds of play. Plutarch speaks of a trained elephant that often practised her steps when she thought no one was looking. No one who has ever visited a zoological park and seen the crowded monkey and baboon cages can have failed to note the wonderful play of these animals. Seals seem never to tire of chasing one another through the water; while even the clumsy hippopotamuses have diving games.

Kittens begin to tumble and play before they are two weeks old. They will roll and toss a ball, hunting it from the dark corners, lay in silent wait for each other, and suddenly spring upon an unsuspecting fellow-cat-baby's back, just as they will do later in life, when seeking their prey. I have seen them play with a catnip mouse for hours at a time, just as the mother cat plays with a real mouse.

Brehm says that this is noticed in their earliest kittenhood, and that the mother cat encourages it in all ways possible, even to becoming a child with her children from love of them, as a human mother does in the nursery with her child. The mother cat begins the play by slowly moving her tail. Gesner considered her tail as the indicator of her moods. The kittens, while they may not understand what this means, are greatly excited by the movement, their eyes sparkle, their ears stand erect, and slowly one after another clutches after the moving tail. Suddenly, one springs over the mother's back, another grabs at her feet, while a third playfully slaps her in the face with his tiny, soft, cushioned paw. She, patiently and mother-like, lovingly submits to all this treatment, as it is only play.

DRYPTOSAURUS. THE PREHISTORIC ANIMALS, TOO, UNDOUBTEDLY HAD THEIR PLAY TIME, WITH GAMES AND "SETTING UP" EXERCISES.

American Museum of Natural History, New York

A HAPPY FAMILY OF POLAR BEARS. THE YOUNG CUBS WRESTLE AND TUMBLE, AS PLAYFULLY AS TWO PUPPIES. THIS PLAY HAS MUCH TO DO WITH THEIR PHYSICAL AND MENTAL DEVELOPMENT.

Many scientists have claimed that this so-called instinct should not be classed as real play. However, such an authority as Darwin thought it was play, and Scheitlin said that the cat let the mouse loose many times in order that she might have the experience of catching it each time. No mercy is shown the helpless mouse, which is the same to her as the toy ball--in the same way as a real beetle and a toy beetle are the same to a small child. Evidently the cat does not play with the mouse for the delight in torturing it, but purely for practice that she may become skilled in the art of catching it. The cat also exercises in springing movements, and by studying the mouse's probable movements, learns to acquire a knowledge and skill in mouse-ways otherwise impossible.

The same cruel practice is found among leopards, panthers, and wild cats. Brehm verifies the observation that many members of the cat family practise torturing their victims in a horrible manner, pretending to liberate them, until the poor creatures at last die from their wounds. Lenz tells of a marten that would play with its prey for hours when not hungry. Especially was this true when marmots chanced to be his victims, and around these he would leap and spring, dealing them terrific blows first with one paw and then with the other. When hungry, however, he proceeded differently, devouring them at once from teeth to tail.

All the cat family, it seems, are fond of human companionship, and take almost as much delight in playing with human beings as with their own kind. This is especially true of the puma. Brehm tells of a tame one that delighted in hiding at the approach of his master and springing out unexpectedly, just as the lion does. Hudson claimed that the puma, with the exception of the monkey, was possibly the most playful of all animals. Travellers tell many interesting tales of the play of these animals, especially on the Pampas of South America.

Gross relates the experience of an Englishman who was compelled to spend the night outdoors on the Pampas of the La Plata. At about nine o'clock, on a bright moonlight night, he saw four pumas coming toward him, two adult animals and two young ones. He well knew that these animals would not attack him, so he quietly waited. In a short time they approached him, chasing one another and playing hide-and-seek like little kittens; and finally leaped directly over the man several times. The mother cat would run ahead, calling to the little ones to follow her. But she never disturbed him.

At times an animal at play with another uses the same tactics and methods employed on its prey. Of course, the value of such practice for the tasks of later-life is evident. Dogs play hide-and-seek, tag, and various chasing games for hours without resting. Among the negroes of the South it is not uncommon to see a hound playing hide-and-seek with the little pickaninnies. I have seen a hound peeping in and out among a pile of brush to discover where the little ones were hiding, and at the first sight of a little black face, he would lay low in anticipation of a playful spring, or a sudden dash-away, with the expectation of being chased by his friends. At times he would suddenly disappear toward his home, and slyly slip around and approach the playground from an opposite direction.

Every one who has owned fox terriers knows how they will crouch in the open grass and remain motionless, with quivering expectation for the other playfellow to arrive, and when the one in ambush sees the other coming he springs toward him, as though he were going to destroy him! And when the two come together, they attempt to seize each other by the necks, as they would do in a real conflict. A wrestle and tussle ensues and when utterly exhausted from this play, the tired dogs, like two fatigued children, run to their homes.

Dogs are fond of playing ball, and will readily bring a ball or stick to their master when he has thrown it. They will also go into the water to bring out sticks that may have been tossed in for amusement. Eugene Zimmerman had a young fox terrier that would set a ball in motion, when there was no one to pitch it for him, by seizing it in his mouth and tossing it up in the air. Monkeys and jaguars will also play ball, and tame bears take great delight in wrestling, playing ball, and fighting mock battles.

American Museum of Natural History, New York

THE MOTHER OPOSSUM IS NEVER HAPPIER THAN WHEN SHE HAS HER LITTLE ONES PLAYING HIDE-AND-SEEK OVER HER BACK.

THIS YOUNG FOX CAME FROM HIS HOME IN THE WOODS DAILY TO PLAY WITH A YOUNG FOX-TERRIER. HE IS NOW RESTING AFTER A ROMP.

Beckmann wonderfully describes the play of a badger, whose only playmate was an exceptionally clever dog, who from his earliest youth had been taught to live with different kinds of animals. "Together they went through a series of gymnastic exercises on pleasant afternoons, and their four-footed friends came from far and near to witness the performance. The essentials of the game were that the badger, roaring and shaking his head like a wild boar, should charge upon the dog, as it stood about fifteen paces off, and strike him in the side with its head; the dog, leaping dexterously entirely over the badger, awaited a second and third attack, and then made his antagonist chase him all round the garden. If the badger managed to snap the dog's hindquarters, an angry tussle ensued, but never resulted in a real fight. If Caspar, the badger, lost his temper, he drew off without turning round, and got up snorting and shaking and with bristling hair, and strutted about like an inflated turkey-cock. After a few moments his hair would smooth down, and with some head-shaking and good-natured grunts the mad play would begin again."

Young animals are strikingly like children in their craving for amusement. A young bear will lie on his back and play with his feet and toes by the hour, while a young pup can have a great game with only a dry bone, or by chasing his shadow on the wall. Rabbits come out in evenings on the sand-hills to play hide-and-seek with their young, and squirrels never weary of this universally popular game. I know of a young fox that used to come from a nearby woods every evening to play with a young fox-terrier. They became great friends and were often seen in the woods together.

A friend who owns a ranch in Texas once raised two young wolves that romped and played with the neighbour's dogs just as if they were dogs themselves. There are other animals, like the weasels, that will also play with strange friends. But they prefer their own kind as playmates. They take the greatest delight in playing with their parents, and nothing is more beautiful or strange than to see several of them playing in a valley on a sunny day. Out pops one little head, with twinkling eyes glancing from side to side, and then as if from nowhere, the little brothers and sisters begin to appear, chasing each other as though they were playing tag. These exercises give them much agility which they will need in later life.

I once owned a tame raccoon, and often kept him chained in the back yard. When he could not find a young chicken or duck to torment, he devised all kinds of schemes to relieve the monotonous hours. He would pile up a number of small stones, and carefully await his chance to fling one into a group of young chickens. He seemed to understand that he was more apt to make a hit when he threw into a crowd than when aiming at a single chick. At other times he would lie on his back, madly waving his tail as though he were signalling for some one to come near. If we chanced to pass by without speaking, he would growl or whine in some way to attract attention. After hours of self-amusement he would lie down as if life were useless, and wait until something or somebody came along to amuse him. His greatest delight was in fishing things out of a pan of water, and he would wash every pebble or plaything that he owned and carefully lay it out to dry. One day he pounced upon a rooster who insulted him by drinking from his water vessel, and plucked a long feather from his tail so quickly that we could hardly realise what had taken place. He then had great fun in attempting to stick the feather in his head or by planting it upright in the ground. Another day, in winter, he broke his chain and made straight for the kitchen, where he found a snug warm place in old Aunt Moriah's kitchen oven. The old negress came to cook dinner and when the raccoon suddenly sprang out of her oven, she vowed, "I'se nevah gwine to cook in dis heah kitchen again; dis place is hoodooed fo' life!"

Once we gave him a pail of hot milk, and it was evidently hotter than we realised; he started to drink it, and suddenly stopped, and in anger grabbed at a very young puppy that was following us, and before we could stop him, dipped the puppy's head into the hot milk. Fortunately, however, the milk was not hot enough to injure the puppy. But the raccoon had taken his revenge out on the little animal, and was evidently satisfied.

It is interesting to note that all animals seem to play games and take exercises that will be especially helpful to them in later life. Badgers, for example, delight in turning somersaults; deer like to jump and leap; foxes and raccoons practise stealing upon one unnoticed; tapirs and crocodiles play in the water as night approaches; mountain goats, sheep, horses and mules run, leap, jump, and play follow-leader. Animals that live in the high mountains practise all kinds of high-jumps, which would be unnecessary if they lived on level ground, but are highly essential in mountainous countries.

Brehm claims that in summer the chamois climb up to the everlasting snow and take much delight in playing in it. They will drop into a crouching position on the top of a very steep mountain, work their four legs with a swimming motion, and slide down on the surface of the snow for a hundred and fifty metres. As they slide down the snow flies over them like a fine powder. As soon as they reach the bottom, they jump to their feet, and slowly climb up the mountain-side again, while many of their comrades silently stand by and watch their coasting approvingly, first one and then another joining in the sport, like human coasters would do. It is not uncommon for a number of them to tumble together at the bottom, like romping children. This coasting is very remarkable, and through skill in it, no doubt, the lives of many chamois are saved from frightful accidents later in life. Alix tells us that dogs of mountainous countries are also often skilled in the art of coasting.

Our tame fawn used to delight in playing with our old rabbit-dog, Nimrod. They were the best of friends, and the fawn would begin the chase by approaching Nimrod as though he were going to stamp him into the earth, and then suddenly leaping quickly and safely over the dog, he would run away. At this signal for a game, if Nimrod was in the mood, he chased the fawn, who would delight in jumping over fences and hedges and waiting for poor Nimrod to get over or under just in time to see his playmate leap to the other side.

Wolves, if taken when quite young, have a most unique way of showing their affection at the appearance of their master. They will spring into the air, tumbling over, with whinnying cries of delight, falling to the ground they pretend to bite and snap at everything, until their friend finally comes very near them.

Prairie dogs are fond of all kinds of races and jumping games; they will each appear at the entrance to their underground homes, and will play a simple form of prisoners'-base for long periods of time. With defiant calls at each other, one finally approaches the home of the other, which is a signal for the third to attempt to slip into the entrance to the second one's home before he can return. Many join in the game and it usually ends in a regular roll-and-tumble for their respective homes.

Perhaps the strangest of all forms of play is that in which young duckbills indulge. They are slightly like puppies in their methods of roll-and-tumble, but the way in which they grab one another with their strange bills, as they strike with their fore-paws is quite original. They seem to have an unusually good disposition, and if one little playfellow falls in the game, and desires to scratch himself before arising, the other patiently waits until he arises, when the mock battle begins anew.

Antelopes have chase and marching games which are beautiful. They seem rapidly to follow an invisible leader over the plains, suddenly forming themselves into pairs, fours, eights, sixteens, until the entire herd thus form one line, like an army of soldiers marching. While this game is progressing, certain of their number stand as sentinels and spectators, and the slightest approach of an enemy is the signal for all play to cease, and for them to disappear over the plains.

When we witness these abundant evidences of the need and prevalence of recreation in the animal world, we are confronted with one more argument for the existence of real mental and moral faculties among our four-footed friends.

IV

ARMOUR-BEARING AND MAIL-CLAD ANIMALS

"The spectacle of Nature is always new, for she is always renewing the spectators. Life is her most exquisite invention; and death is her expert contrivance to get plenty of life."

--GOETHE'S *Aphorisms* (trans. by HUXLEY).

Civilised nations throughout the world at different times in their country's history have protected their soldiers and warriors with coats of armour or mail. This practice prevailed extensively during the Middle Ages; but it has almost entirely disappeared. The German breastplates of to-day are an attempted revival. The coats of mail of the ancient warriors underwent an evolutionary process, until they were indeed brought to a high pitch of perfection and beauty. It was at this period that they were abandoned as too burdensome to be of practical value.

This protective form of armour has been used by animals since time immemorial, and was copied by man from them; and among the various forms of it are found examples of every kind of armour used in the human world, from the rough leather shields of hide which the savages use, to the ornamental suits of mail, like those used by the knights of the fifteenth century. Indeed, some animals have carried the art of protection to such an extent that they are veritable movable forts, or "tanks!"

In the early part of the earth's history, animals needed greater protection from powerful enemies than they do at present, and they developed a coat of mail, exquisite in appearance and even more efficient than that used by man. Yet, like mankind, they have found newer and more efficient methods of protection, and as a result of changed conditions and enemies, have discarded, at least most of them, their coats of mail and armour. Most of those who have held to the old-fashioned ways of fighting and facing the world, have, like unprogressive peoples, perished; and to-day only a few armour-bearing animals exist. These classes, however, have never been very large, and consist of two small families; the pangolins and the armadillos. The former live in southern Asia and Africa, while the latter are inhabitants of South America.

These animals have a great advantage over man, for their armour grows upon their bodies and is a part of them, while man must put his on and take it off and continually replace the worn-out parts. Again, while there are only three distinct kinds of human armour--the chain, scale and plate armour--there are many kinds of animal armour. What wonderful opportunities exist to-day in the great museums for studying the different kinds of animal armour, for those who are interested!

The scaly ant-eater, who is at home in Africa and Asia, is one of the most unusual and original types of mail-clad animals. He might be compared to a wolf in outline, covered from head to tail in huge, horny plates, which look like immense finger-nails overlapping each other. His head sharpens out into a long, narrow snout, which contains a sticky, worm-like tongue, and this he can use with great rapidity and effect in raiding an ant-hill. He drops his tongue over the entrance, and the ants attempt to crawl over it and are glued to it. He walks in a very unique way by going upon the backs of his feet. This preserves his wonderful claws for bursting open ants' nests, as his chief food consists of these tiny insects and their eggs.

A cousin of the scaly ant-eater, the great ant-eater of South America, has the same general habits of his near-kinsman. He has an immense bushy tail with which some naturalists claim he sweeps up ants. This is not true, however; he uses his tail, when he lies down, to cover himself. The hairs of the tail part in such a manner as to fall over the body like a thatched roof, protecting it from rain and storm alike.

A part of the head and under portion of this ant-eater's body are unprotected, and this is why he rolls himself up like a ball when danger is near. In this position, his scales stand out in such a way as to make a complete row of sharp points, as uninviting as the wires on a barbed wire fence. Yet, it is claimed that certain of his enemies, like the leopard, know his one great weakness--a terror of being wet--and often make him uncoil by rolling him into the water. His coat of hard covering is really compact masses of hardened hair drawn out to sharp dagger points, and might be likened to pine cones endued with power. Through ages of experience, the scaly ant-eater has learned that even his powerful coat of protection is not altogether a success in life's battles, and from time to time his armour has been made lighter and lighter, and because he has been so slow in making the necessary changes, he is to-day very scarce, and able only by the greatest caution to drag out a dull existence as a nocturnal and burrowing animal. It would seem that with such powerful protection as he originally had, he would have outlived the puny armadillos, but his fast disappearance proves that the race is not always to the swift, nor the battle to the strong.

Among the animals which have discarded their old-fashioned coats of mail, and have successfully protected themselves against all enemies, may be mentioned the frogs, newts, and their kinspeople, the reptiles. These latter, the learned, with their delight in multiplying terms, have classed as amphibians. During the period when the coal forests were growing over what we now know as England, there were innumerable amphibians, and even to-day their petrified footmarks are found in sandstone. The underside of their chests were covered with large bony plates, and in some cases the rest of the body was covered with scale-like bones. Yet, all the newts and frogs of to-day have wisely discarded the old coats of armour used by their forefathers.

The armadillo has an armour of quite another kind, notwithstanding the fact that pangolins and armadillos belong to the same great family, and each eats ants. Their plates of armour, or shields, have nothing at all to do with the hair, nor do they have anything to do with the exo-skeleton; they are formed of bone material, which appears in the true skin in the form of tiny shields, and each shield is itself covered with a hard plate which grows in the outer skin. The actual formation of these shields differs largely in the various species of armadillo.

American Museum of Natural History, New York

NAOSAURUS AND DIMETRODON, TWO EXTINCT ARMOUR-BEARERS WHO SHOULD HAVE BEEN WELL ABLE TO PROTECT THEMSELVES.

AN ARMOUR-BEARER OF PREHISTORIC TIMES WHOSE SHIELD WAS AN EFFECTIVE PROTECTION AGAINST ENEMY HORNS.

It is well to remember that the pangolins and armadillos are the last survivors of a great and ancient family of armour-bearers. Many of their remote ancestors have been found in the rocks and hills of South America, and all of their representatives of to-day are small animals--the last of a doomed race--creatures of yesterday. The glyptodon is known to have been more than eleven feet in length, and his near-kinsman, the chlamydothere, was even larger. He was nearly the size of our present-day rhinoceros. These extinct giants carried on their backs huge domes of bony plates, that must have rivalled our much-feared tanks, of trench war fame. One would think they were invulnerable, yet the glyptodon and the chlamydothere, with many other equally well protected creatures, have long ago disappeared from the earth, but how and why nobody knows. This total disappearance of these marvellously protected giants, which seemed capable of defending themselves against any and all kinds of enemies that might have arisen, is one of the strangest and most unsolvable problems of science.

Another mail-clad animal of importance is the armadillo of the tropical and temperate regions of South America. He is nocturnal in habits, sleeping in his underground home during the day, and coming out at night to seek for food. This underground home is rather large, and the nursery is well protected from enemies by its location. In it the mother armadillo rears her young until they are large enough to care for themselves.

All species of the armadillos are powerful burrowers, and they are well equipped for their tunnelling in the earth with strong fore limbs. They feed upon all kinds of insects and animal substances. It is claimed that the giant armadillo is a veritable grave-robber and sometimes digs up dead bodies for the purpose of eating them.

These animals are plentiful upon the savannas of South America, and they feast upon the bodies of dead cattle. So hard are their coats of armour that the Gauchos sharpen their Spanish knives, which they always carry, upon them. Should the armadillo be attacked by a man on horseback, he will burrow so rapidly that only by the quickest movements of the man can he be caught; and if he is, watch out for his terrible claws!

No animal is better protected by nature from its enemies than the pichiciago, whose scientific name is *chlamyphorus truncatus*. This strange little mantle-bearer wears a coat of mail which is as flexible as the human-made coats of armour of olden times, and he is as safe under its cover, which allows him perfect freedom, as if he were under the ground. He is about the size of the ordinary mole, and his general habits are not unlike those of the mole. He is an underground-dweller, with enormous fore-paws, palm-shaped, upon which are five powerful claws. These he uses to great advantage in digging in the earth for insects and for building his home. He has a small snout, reminding one of that of a pig; while his piercing little eyes are deeply hidden in his fur. He is a native of Chile, and because of his shy nature and subterranean habits is rarely seen.

The most interesting feature about this little creature is the cuirass which so perfectly protects his body. Its formation and arrangement is quite unusual; it appears like a number of squared plates of horn, tightly united to short strips of tape, which are sewed together. The cuirass is not connected with the entire body of the animal, but only on the top of the head and along the spine. It covers the entire back, and when it reaches the tail, turns downward, forming a perfect flap, which protects the hindquarters.

The various species of manis are famed for their powerful coats of armour. They, also, belong to the great group of burrowers, and their coats of mail assume both offensive and defensive characters. These mail-bearers are covered with numerous sharp-edged scales, like miniature horns, which entirely overlap one another, like shingles on a house. They are of great hardness, and form a belt which no animal of their regions can penetrate. A revolver shot will produce not the slightest effect upon the body of this iron-protected animal.

These animals are plentiful in India, and when they are molested, they deliberately wind themselves up, coil their tails over their bodies, and remain in conscious security against the fruitless blows of their enemies, who soon weary of the wounds caused from the prickly scales of impenetrable armour.

Instead of wearing heavy coats of mail, certain animals, such as the hedgehog and porcupine, prefer to wear coats covered with needles and pins. Of course, a coat of spines is used purely for protection. And against the attacks of such enemies as dogs, it proves all-sufficient, but it is a well-known fact that pumas and leopards will kill and eat porcupines at all times, paying small attention to their spines, as is shown by the number which are sometimes found sticking in the body of a porcupine-eating animal.

There are several species of this great spine-bearing family; and many of them, especially the true porcupines and the echidnas, have burrows in the ground and thus have a double means of protecting themselves. But others, such as the hedgehog, depend for their protection upon their ability to roll up into a ball, thus presenting a barbed wire protection. Still others live largely in the trees and seek by other means to protect themselves.

One of the most interesting coats of armour is that worn by the porcupine ant-eater--oft-times erroneously called porcupine or hedgehog. He is a native of Australia, and is a powerful burrower. He is marvellously protected by means of a coat of needles or spines which inflict painful wounds on the dog or other enemy that ventures to attack him. In case of danger, he curls himself up into a ball, and defies any one to come near. Not only does he possess the coat of prickles with which he defends himself, but he also has a large perforated claw or spur on each hind foot through which pours an ill-smelling liquid, and these also aid in protecting him. There are several varieties of porcupines which inhabit Asia, Africa, Southern Europe and America.

When a porcupine wishes to attack an enemy, he rushes at it backwards, and usually leaves the enemy literally covered, like a living pin-cushion, with his spines. These animals have convex skulls, short tails, and live chiefly in the warmer regions of the Old World. Those of America are different in one particular--the soles of their feet are covered with hard, bone-like tubercles, instead of being soft and smooth; there are also a number of hairs that are intermingled with the spines. The Canada porcupine has more hairs than the American, and a shorter and stumpier tail.

Another animal whose methods of defence are by means of his spines, is the hedgehog. His spines do not terminate in sharp points, like those of the porcupine, but end in tiny knobs. These are placed beneath the skin, and are like pins stuck through a cushion. The hedgehog, like the porcupine, rolls himself into a ball when attacked by enemies, and he has the additional ability of throwing himself down a hillside, like a rolling ball, and thus escaping his enemies without injury to himself. It would seem that the hedgehog, rolled into a ball and covered with prickles, would be protected from all enemies. But this is not true, for the clever fox knows just how to make him unroll. This one secret of the hedgehog's weakness very often causes his loss of life. His weakness is a terror of being wet or dropped into water; and when the fox finds him all rolled up, he carefully rolls him into a pond of water and, when he unrolls, quickly drowns him. Notwithstanding the shortness of the hedgehog's spines, he is the most highly specialised of all spine-bearing animals. In the lower order of animals there are spiny mice and spiny rats, and even the horned toad uses his horns as a means of protection against his enemies.

One of the most peculiarly armoured animals is the horned lizard, commonly known as the "horned toad" of America. His body is covered with small spiny scales, while the chisel-shaped head has a circlet of miniature horns. These he uses when attacked by enemies to shield himself against bites and knocks. The Indians claim that if a snake swallows the horned lizard whole, the lizard will immediately work his way through the snake. This would not be without a parallel, however, for it is generally known that box-fishes, when swallowed by sharks, bite their way out!

Nature has been especially kind to horned lizards, and that is the reason there are so many of them. They well know the secret of the Gyges ring, and can put on the garment of invisibility in a very short time. They especially frequent the desert regions of the South and West; and those that dwell in black sandy regions are black; those of red clay regions are red; those of grey regions, grey; those from the variously coloured regions of blue and red are precisely the colour of the earth. But not satisfied with all their protections of armour and camouflage, they actually, when hard-pressed by an enemy, feign death, like an opossum! And if the enemy persists in his attack, and Mr. Lizard cannot escape, as a final effort he spurts tears of blood from his eyes. The Mexicans call him the "sacred toad." The phenomenon of blood-shooting has been explained in various ways, all of which seem equally unsatisfactory. So far it is one of Nature's secrets. Perhaps some day we may understand it.

The tortoises are among the best examples of creatures which to-day protect themselves with armour. They are, of course, reptiles, yet in the general formation of their armour, they are strikingly like armadillos. The tortoise has his armour so arranged over his body that it forms one big box. He draws his head and limbs into this whenever danger is near. In Texas recently I found a small land terrapin, and as soon as I came near, he closed his house. I picked him up, and then carefully laid him upside down on the ground, and stepped behind some nearby bushes to see what he would do. Immediately he poked his head out, and then his feet, and then he began to wave his feet wildly in air, and finally threw himself in the right position and hastened away through the grass.

The turtle protects himself in the same way, and draws his head, feet, and tail under his own house-roof where nothing can get him.

Lobsters and crabs are excellent types of armour-bearing animals. Lobsters wear marvellous coats of mail, very similar to those worn by human warriors during the age of chivalry. Their jointed structure assures them perfect ease and security. Crabs, however, believe, as the tortoise, in the strong-box protection. When resting, crabs tuck their legs beneath them, so as to shelter themselves under the hard covering. Upon crabs Nature has bestowed twin protective characteristics: namely, they are armoured, and also mimic their surroundings. The latter protection is especially needful, because certain big fishes, like the cod, are in the habit of swallowing crabs whole. In this case the armour is of no use, while the protective resemblance saves the crab.

To discuss in detail all the various kinds of armour and mail that the different groups of animals have used and developed for offensive and defensive purposes since the days of the prehistoric gigantic armadillos to the present, would require a book of itself. It is sufficient to know that armour and mail and spines are among Nature's most common forms of protection, and that each age develops new and ever more efficient methods of defence. This simply means that the age-long drama of evolution is always changing. Everything that is came out of that which was, and throughout the ages the ever-evolving organisms have been developing out of the past, that they might ever be new.

V

MINERS AND EXCAVATORS

"When the cold winter comes and the water plants die,
And the little brooks yield no further supply,
Down in his burrow he cosily creeps,
And quietly through the long winter sleeps."

--*(The Water Rat.)*

There are many ground-dwellers in the animal world, and foremost among them is the mole. This remarkable little creature is not only gifted as a digger of canals and tunnels, but plans and makes the most extraordinary subterranean homes. Sometimes he unites with his fellow creatures and establishes whole cities with winding passages, chambers, exits and entrances. In fact, he has not only an exquisitely arranged home, but highways and roads that lead to his kingly hunting-grounds which are as elaborate as that of a modern man of wealth and culture. Indeed his subterranean network of tunnels excels in complexity our modern city subways. His engineering calculations never fail, and a cave-in of his hallways is unknown. This little gentleman with the velvet coat is a genius of varied accomplishments!

But this is only true when the mole is in his proper sphere or home. There he can fight like a tiger, catch his prey both below and above ground, build wells to collect and retain water, swim like a fish, and do many things which would seem impossible, judging from his awkward and clumsy manner above ground.

His apparent awkwardness while out of his natural habitat is largely due to the peculiar formation of his limbs, and the stupid appearance of his small half-hidden eyes. These features seem to mark him to the casual observer as a dull animal, yet in reality he is very active and bright, and when at home displays his marvellous genius in many ways! His upturned hands become powerful shovels, and by the aid of an extra bone, the sickle, which belongs to the inside of the thumb, he is enabled to work like an athlete. His velvet-like hair stands straight up, like the pile on velvet, and his tiny eyes are so hidden by hair that they do not get injured. The eyes are not well finished from an optician's point of view--but they serve admirably all the needs of the mole's life. As dull and stupid as he appears, he is, considering his size, the fiercest and most active animal in existence. Imagine him the size of a wild cat! He would be a beast of exceeding ferocity. Even a lion would find him a formidable antagonist. With such an animal tunnelling in his fields and cellars, man would have a terror hard to exterminate.

The mole is an engineer and miner who seems to have a strange sense of direction practically unknown to many other animals. How he manages to form tunnels and burrows in lines of such unusual straightness is unknown; he always works in darkness, unless it is that he can see in the dark. His little hills are not deliberate structures; they are only shaft ends through which this miner throws out the earth that he has scooped from subterranean depths, and in most cases smoothed out so that if an observer examines the burrow he will find only solid earth, and a road into his tunnel which leads to his real habitation.

The home of the mole is usually beneath a tree or hillock, and reminds one of a miniature city of tunnels and engineering feats. The main, or central, room is shaped like a great dome, the upper part of which is level with the earth around the hill, and therefore nowhere near its apex. Mr. Wood has verified the observation that around the keep are two circular passages, one of which is level with the ceiling, while the other is above. The upper circle is decidedly smaller than the lower; and there are five ascending passages which connect the galleries with each other. There is only one entrance, however, and from it three roads lead into the upper part of the keep. When a mole enters the house from one of the tunnels, he must go through the basement in order to get to the upper part of the house and so descend into the keep. There is still another entrance into the keep from below. One passage leads downward directly from the middle of the chamber, then curving upward, leads into a larger tunnel or subway.

Throughout the vast network of tunnels every inch of wall space seems quite smooth and polished. This is due to the continuous pressure of the mole's fur against the walls. Thus there is little danger of the walls collapsing even after a rain-storm. No human being knows just why the mole has such a complex system of underground streets and tunnels; perhaps it is because he finds that a greater feeling of safety surrounds his home when he knows that in case of danger he can escape in a dozen directions. Surely he is the original builder of labyrinths!

How marvellous that so tiny a creature can build such a fortress! The complex chambers and circular galleries do justice to an artist. The space of ground covered by a single mole's roads and galleries is almost unbelievable; in every direction from the fortress they run, and are sunk at various depths, according to the condition of the mole's hunting-grounds, which are really the spaces of ground through which he tunnels. Worms and underground insects are his chief food. Sometimes he ploughs along the surface of the ground, and exposes his back as he works; but if the weather is dry, he ploughs deeply into the earth for worms. He fills his storehouse with earth-worms for winter use, and he finds it necessary to bite their heads off, which leaves them inert but not dead. This cannot be done in the summer months without the heads regrowing and the worms crawling away. The mole knows the exact temperature best suited for keeping his meat fresh!

A most interesting and beautiful family of miner-cousins of the moles are the shrews. They are excavators of great ability, and because of their nocturnal habits are rarely seen alive. They are very similar to the mole, though much more handsome. Their domicile is built of dry grass at the end of a tunnel.

The shrew mole of North America is a ground-digger of great ingenuity. He is second only to the mole in the extent and pretensions of his engineering and tunnelling. His eyes are very small and deeply hidden in his fur. During the day he constantly comes to the surface of the earth, and one may catch him by driving a hoe or spade underneath him.

Another underground-dweller is the elephant shrew of South America. He has a long nose, thick fur, short ears, and, unlike his cousins, he loves to bask in the warm sunshine. At the least signal of alarm he darts away to his subterranean home. As a mining engineer he is unexcelled; he sinks his tunnels by first boring an almost perpendicular shaft, and then making his burrows at an angle. It is a sad day for earth-worms when he decides to locate in their vicinity!

It is not an easy task to classify the homes of animals. Many of them have characteristics that entitle them to be placed under several groupings. The otter, for example, might be classed as a cave-dweller, as he seeks refuge in caves; yet he also rears his young in underground nests as a burrowing animal. But few naturalists believe that he does his own digging. This is not surprising when we remember that there are many other animals that live in caves and grottoes, and like the otter, seek ready-made homes for their convenience. Among these may be mentioned three American salamanders, bats, and a few strange mice, who seek darkness and constant temperature, and therefore find caves best suited to their needs.

The same is true of the weasel, who is thought to be a great burrower, but in fact, like our remote cave-dwelling ancestors, makes his home only in caves, in rocky crevices, and under the gnarled roots of old trees. He is a bright-eyed little creature, with a slender snake-like neck and red body. He is a great friend of mankind, as he does more toward eradicating mice and other nocturnal depredators than all the rat-catchers in the land. His home is quite ordinary compared to that of the more ambitious underground-dwellers.

A near cousin of the weasel, and a most ingenious engineer and miner, is the badger. He is a tenement-dweller and builds his home in the deep, shady woods. His home is rather pretentious with several chambers, and a most delightfully furnished nursery which is warmly padded with dry grass and moss.

The badger, once so plentiful in England and America, is fast passing away because of the increase of towns and cities. As soon as the forest in which he dwells is drained and converted into farm land, the badger disappears. He is driven from the soil where he once held sway, and is one of those unfortunate animals which are eliminated by man-made civilisation.

The fox of the Far North is a famous excavator, and his underground home which shelters and protects him from the extreme cold is most spacious. It is a strange fact that these cunning little animals rarely make their homes away from others of their kind. Sometimes twenty to thirty are found in close proximity. And their owners are unquestionably the smartest, keenest, and quickest creatures that roam the wilds. While some of their deeds are questionable, their quick wits and nimble bodies excite our admiration.

These arctic foxes really build small cities, and their semi-social life may be accounted for by the peculiar suitability of the place which they select for a habitation. Their homes are usually in a sandy hill, where it is very easy for them to burrow; and the strangest part of the whole city is that each burrow is complete and entirely independent in itself. There are many winding paths and tunnels in each house, but each belongs exclusively to its owner and never winds into a neighbour's house. In case of danger the fox has many directions in which he may escape.

The nursery is the most carefully arranged of all the rooms. It is rather small and is directly connected with the main outer chamber somewhat like the nursery of the mole. So skilfully is it situated that it sometimes happens a hunter will dig into a fox's burrow and never discover the nest of young, and later the clever mother will return to carry away her babes, which are usually five to six in number. Adjoining the nursery are two or three storage rooms filled with food for the winter. The number of bones usually found in the basement indicates that a great variety of ducks, fish, hares, lemming, and stoats are regularly eaten, and that the average fox family does not want for food.

The arctic fox is not only a beauty in his coat of pure white, but is unusually brainy. Persecuted animals, like persecuted human beings, become very wise. Nature is kind to the fox in his arctic home, and in the winter turns his coat snow white so that he may easily escape his enemies--especially men, who seek his beautiful fur and edible body. He is skilled in his distrust of wires, sticks, guns and strings! No man knows better than he the meaning of foot-tracks in the snow, and how long they have been there, and which way they lead; thus, those that survive their enemies have acquired extreme wisdom, and keep carefully away from everything that is at all suspicious to their eyes and nostrils.

The Siberian fox is one of those wise creatures that has defied in a most extraordinary way his handicaps, and, refusing to admit them, has boldly selected the strangest dwelling-place known to the animal world--the horn of the mountain sheep. This unique dwelling-place has been the home of the Siberian fox for ages, and his ancestors have known no other. The mountain sheep, which are giants among their kind, have the longest horns in proportion to their size of any animal in existence. The argali of Siberia is the largest of all sheep, and is equal in bulk and weight to an average-sized ox, with horns proportionally large. The horns of these animals are strikingly like those of the Rocky Mountain sheep of America, except they are much larger. They spring up from the forehead, tilt backward, then boldly curve below the muzzle, before finally again pointing upward and tapering into a sharp and delicate point. They are hollow, though exceedingly stout and elastic, and strengthened on the outside by a number of ridges or horny rings set very close together. They are found in large numbers in this land of perpetual ice and snow, and it is thought that they break from the sheep's heads very easily.

It is not uncommon to find them lying in a spot which has been a battlefield, where two sheep in attempting to settle some dispute have fought and fallen. It is not long after they have thus fallen before they are utilised by Mr. Fox. He stores himself carefully away in these roomy horns, one of which Mrs. Fox uses as a nursery, finding it a snug, safe, and warm place to rear her little family.

The other varieties of foxes, especially the grey and red, are not so skilled in home-making. This may be due to the fact that they do not have need of such elaborate houses as their arctic cousins. Again, it may be that the existence of numerous deserted homes of badgers, or even rabbits, makes it unnecessary for them to spend their time in building homes of their own. It is much easier to enlarge the ready-made burrow of a rabbit than to dig a new tunnel, of course.

If there is no ready-made burrow to be had, then the wise fox sets to work and scoops out his own. Herein he sleeps all the day, and comes forth only at night. A small chamber from the main room serves as the nursery, and here the babies are born and nurtured. Nothing is more beautiful than to see the entire family--mother, father, and children--come forth at evening to play. The young are as sportive as pups, but they never wander far from home. Their broad heads, grey coats, short tails and awkward appearance would lead no one to think that they were the children of handsome, nimble-limbed, intelligent Mrs. Fox!

Woe to the dog that enters Mrs. Fox's home! She is a pugilist of the first order, and knows how to fight far better than the average bull terrier. It requires a very savage dog to kill her, and he is apt to be minus an ear when the battle is over.

Red and grey foxes are similar in intelligence, but differ in many other ways: the former are like the gipsies in always moving about from place to place, while the latter stick to one general locality, although their hunting-grounds may range for several miles in all directions. Red foxes seem actually to enjoy being hunted by dogs; in most cases they will outrun the dogs, and rarely seek protection from caves or rocks.

The grey fox, on the other hand, cares little for racing, but seeks protection among rocky cliffs where the dogs are at a disadvantage. Here none but the smallest canines may enter the holes and crannies, and they are usually wise enough to stay out. Hunters are thoroughly familiar with the tactics of the fox family, and therefore select the red ones for their sport.

The foxes are truly famed for their cunning, and when other animals try to play tricks on them, the trick usually turns out in the foxes' favour. During the winter season these wise creatures are sometimes hard pressed for food. Birds and small animals are hard to catch, and the farmers' chicken houses are closed. It is then that the wise fox needs all his wit and wisdom, for he oftentimes becomes the hunted as well as the hunter. His chief enemies are the puma and the timber wolf, but they are seldom able to get him.

The prairie-dog is so talented that he might be classed under several headings; he is sociable, a burrower, and especially gifted in the art of constructing underground "dog towns." He is rarely called by his Indian name, *Wish-ton-wish*, and we know him only as the prairie-dog. Evidently he was given this name because of his yelping bark, which resembles the cry of a young domestic dog.

He is a good-looking but rather curious little animal. He has a round, flat head, and garish-red fur, and a stout little body. He makes an affectionate pet, and loves the society of human beings. When he decides to start a town, he usually succeeds, for he is an exceedingly prolific animal, and his extensive burrows seem to have no ends. They are rather large, and run to great depths. In the western part of the United States, especially on the big prairies, the prairie-dog towns often cover large areas. They are usually dug in a sloping direction, and descend four to six feet in depth, and then suddenly rise upward again. Hundreds of these little tunnels are dug in such close proximity to each other that it is quite unsafe for cattle and horses to pass over them. This is the chief reason why ranchmen do not like the otherwise harmless little animals of the prairies.

These dog towns are most curious, and a visit to one of them well repays the traveller. Strangely enough, the prairie-dog is exceedingly inquisitive and this very quality often costs the little animal his life. Mr. Wood, in describing the prairie-dog's habits, says that this wise little Westerner, when perched on the hillocks which we have already described, is able to survey a wide extent of territory and as soon as he sees a visitor, he gives a loud yelp of alarm, and dives into his burrow, his tiny feet knocking together with a ludicrous flourish as he disappears. In every direction similar scenes are enacted. The warning cry has been heard, and immediately every dog within a hundred yards repeats the cry and leaps into his burrow. Their curiosity, however, cannot be suppressed, and no sooner have they vanished from sight than their heads are seen protruding from their burrows. Sometimes hundreds of them will be peeping from their homes at one time, their beautiful eyes sparkling as they cautiously watch the enemy's every movement.

The prairie-dog is truly a tenement dweller, and his home is occupied not only by his own kind, but by owls and rattlesnakes. Most naturalists believe that these incongruous families live in perfect harmony; but it is a well-known fact that the snake occasionally devours the young prairie-dogs, and he must be considered by them as an intruder who procured board and lodging without their consent. The owls, on the other hand, are supposed to do no harm, although it may be that they also occasionally feast on a tender young pup.

The magnificent little animals known to scientists as vizcachas, and whose homes are on the pampas of South America, are the most skilled builders of underground cities in the animal world. Their villages or cities are called "vizcacheras" and are provided with from ten to twenty mouths or subway entrances, with one entrance often serving for several holes. If the ground is soft, it is not uncommon to find twenty to thirty burrows in a vizcachera; but if the ground is rocky and hard, only four or five burrows are found. These wide-mouthed, gaping burrows are dug close together, and the entire town usually covers from one hundred to two hundred square feet.

The vizcacheras are different from other underground animal cities; some of the burrows are large, others are small. Most of them open into a subterranean main-street at from four to six feet from the entrance; from this street other streets wind and turn in all directions, like a man-made subway, and many of them extend clear into other streets or subways, thus forming a complete network of underground passageways. All the tunnelled-out dirt is brought to the surface and forms a large mound to prevent the water from entering the cities.

According to W. H. Hudson, in *The Naturalist in La Plata*, "in some directions a person might ride five hundred miles and never advance half a mile without seeing one or more of them. In districts where, as far as the eye can see, the plains are as level and smooth as a bowling-green, especially in winter when the grass is close-cropped, and where the rough giant-thistle has not sprung up, these mounds appear like brown or dark spots on a green surface. They are the only irregularities that occur to catch the eye, and consequently form an important feature in the scenery. In some places they are so near together that a person on horseback may count a hundred of them from one point of view."

Unlike some burrowing animals, the vizcacha does not select a spot where there is a bank or depression in the soil, or roots of trees, or even tall grass; knowing that they only attract the opossum, skunk, armadillo, and weasel, he chooses an open level plot of ground where he can watch in all directions for enemies while he works.

The great or main entrance to some of these underground cities is sometimes four to six feet in diameter. A small man stands shoulder deep in them. The going and coming of these little vizcachas would almost lead one to believe that they have a primitive city government, and are ruled according to definite laws. Their cities stand for generations, and many of the old human inhabitants tell of certain vizcacheras around them which existed when their parents were living. The founder of a new village is usually a male; and he goes only a short distance from the other villages to establish his new colony.

These cities are by no means occupied by their builders alone, but have their undesirables within their borders. The unique style of burrowing which the vizcachas employ benefits several kinds of birds, especially the Minerva, and one species of the swallows, which build their nests in the bank-like holes in the sides of the vizcacha's cities. Several insects, among which may be mentioned a large nocturnal bug, with red wings and shiny black body, also seek the same shelter; another foreign inhabitant is a night-roaming cincindela, with dark green wing-cases and pale red legs, which remind one of oriental jewels. There are also no less than six species of wingless wasps, beautifully coloured in red, black, and white. Dozens of spiders and smaller insects that live in and near the vizcacheras, which are everywhere sprinkled over the pampas, pass in and out among the streets recognising their respective friends and enemies.

The home life in these communities is most interesting. The burrowers remain indoors until late in the evening during the winter, but in summer appear before the sun sets. One of the larger males is the first to appear, as if to see if everything is safe from danger; if it is, others immediately pop up and take their places at the entrance to the burrow. The females are smaller than the males, and stand up that they may see everything that happens. Curiosity struggling within them for mastery is often the cause of their death. Tiny swallows hover over the entrances, like myriads of large moths, with never-ending low, mournful cries.

Of all the incongruous inhabitants of the vizcacheras, the fox is the most dreaded and the least welcome. To appease his growls and snarls the vizcachas are sometimes forced to let him occupy one of their rooms for a season, or even permanently. During a part of the year he appears quite unassuming and indifferent to the general affairs of the household, and he really goes quite unnoticed, even though he may be sitting on the mound in the family group. But when the vizcachas appear in the spring, the fox begins to become interested in the nursery and as soon as the older animals are away he devours the young. Occasionally, if the fox is hungry, or if he has another friend to aid him, he will hunt the vizcachera from end to end, battling with the old, and usually killing all the young. It often happens that the mother vizcacha, when her babes are large enough to follow her, will take them away to another place that is safer.

The language of these city-builders is most unusual; the males frequently utter the most varied and astonishing cries. They are jarring in the extreme, and are produced in the most leisurely manner, growing louder and louder and finally ending with a slow quaver. At other times, they grunt like small pigs. Hudson says that any quick noise, like the report of a gun, produces a most startling effect among these little animals. As soon as the report is broken on the stillness of the night a perfect furore of cries issues forth from every direction. In a few seconds it ceases for a momentary lull, and then suddenly breaks forth again, louder than before. The tones of the different ones are so different that the cries of nearby individuals may be plainly distinguished amidst the babel of voices coming from the distance. It sounds as if thousands upon thousands of them were striving to express every emotion with their tiny tenor voices. No words can describe the effect that these sounds produce. One of the most peculiar calls is the special alarm-note, which is sharp, sudden, and shrill. It is reported from one to another until every vizcacha is safe in his burrow.

But with all the kind and sociable qualities of these little animals, they have characteristics which seem rather paradoxical, and chief among these is their resentment of any intrusion of neighbours into their burrows. Although a number of individuals may reside in adjoining compartments in the same burrow, yet if one enters a burrow not his own--woe is he! Even when pursued by fierce dogs a vizcacha will rarely enter a room of another. If he does, he is immediately pounced upon by the angry owner, and is usually driven clear out of the burrow. These animals are undoubtedly far the most versatile and intelligent rodents in the world.

A most unusual miner and underground dweller is the pocket gopher of North and Central America. He is a rat-like animal, and is most plentiful on the plains of the Mississippi region. He is unusual in appearance, dressed in brown and grey fur, with tiny white feet, small eyes and ears, and a short stubby tail. His feet are wonderfully strong, and his fore-paws are armed with strong, curved claws. But he is famed for his wonderful fur-lined pouches which open inside his cheeks and serve a peculiar use.

His entire life, with rare exceptions, is spent underground. There he makes long tunnels for the purpose of securing tender roots for food; these tunnels are about twelve to eighteen inches below the surface, and usually wind under the foot of a tree where a sinking passage goes down four to five feet further and leads to a large living-room. This is the family nest and nursery, lined with grass and soft fur which Mrs. Gopher has taken from her own body. Adjoining the living-room is a storage bin filled with nuts, dried bits of roots, tobacco, and potatoes.

Much that is exaggerated has been said in regard to the adaptability of the gopher for his work. But it is a fact that he is of all the diggers best suited for his task. He uses his strong teeth, like a trench-digger uses a pick, to loosen the earth; and while his fore-feet are kept constantly at work in digging and pressing the dirt back under the body, the hind feet also aid in shovelling it still farther back. When a sufficient amount has heaped up behind him, he performs the strangest of all his feats--he turns around, and places his hands vertically against his chin, thus forcing himself backwards, pushing the dirt ahead of himself until it is forced out of the tunnel. At the outer end of the tunnel is formed a little hillock.

Dr. Merriam has made a special study of the gopher, and in speaking of the strange habit of running backwards, he says that even in carrying food to one of his barns or storehouses the gopher rarely turns round but usually runs backwards and forwards, over and over again like a shuttle on its track.

The gopher uses his pouches for carrying food, not dirt. When he has eaten a sufficient amount of food, he fills his pouches. If a potato is too large to be carried in this way, he trims it off to the right size. His method of emptying his pouches is most interesting; with his two tiny paws he delicately presses the food from his cheeks.

The woodchuck is an American basement-dweller of considerable renown. His peculiar whistling cry has won for him from the French the name of *siffleur*; and we sometimes call him by the very inappropriate name of ground-hog. He is a skilled weather prophet, and his appearance in the early spring signifies that the winter is over. He never shows himself until the cold is gone.

The home of the woodchuck is usually found under a hill, with a sheltering rock to protect the entrance, which leads into a tunnel, from twenty to thirty feet in length, finally ending by entering his home proper. The tunnel descends obliquely for several feet, and again rises towards the surface. His nest is rather large, and nicely lined with dry grass and leaves, which serve as a carpet for the young woodchucks when they come into the world. The young remain in the underground home until they are about five months old, then they go out into the world for themselves.

The ground squirrel long ago decided that he would rather have a dwelling under the ground than in the tree-tops, for in an underground home he would have more protection, a better place for storing food, and a far safer nursery for rearing his precious babes. So snug, cosy and hidden are the tiny quarters to which his runs or subways lead that his family is quite safe against most enemies. The ingenuity and skill shown in the construction of his home entitles him to rank among the leading animal miners and excavators.

The most unusual of all the underground and basement dwellers is the polar bear. This wise inhabitant of the Far North has long ago learned that no animal needs to freeze to death in the snow. To him the snow is a constant means of warmth and protection, and as winter approaches, he seeks a position, usually near a big rock, where he digs out a hole of small dimensions, and allows the snow to cover his body. Strangely enough it is only the female bear that seeks this permanent snow hut; the males do not care to spend so much time in seclusion. The same is true of the unmated females. But the mated females always have snow huts in which they give birth to their young, and where they reside until early spring; then the mother bear comes forth with them to seek food and teach them the ways of the world.

American Museum of Natural History, New York

TO THE POLAR BEAR THE ICE AND SNOW OF THE FAR NORTH MEAN WARMTH AND PROTECTION. THE MOTHER BEAR DIGS HERSELF INTO A SNOWBANK, WHERE SHE LIVES QUITE COMFORTABLY THROUGHOUT THE WINTER.

THE SHARP CLAWS OF THE GROUND SQUIRREL ARE EFFICACIOUS TOOLS IN DIGGING HIS COSY UNDERGROUND BURROW.

There is no danger that the bears will stifle for air under the snow, because the warmth of their breath always keeps a small hole open at the top of the snow-cell. This snow-house increases as time goes on, the heat exhaled from their bodies gradually melting the snow. Often Mrs. Bear's home is discovered by means of the tiny hole in the roof around which is collected quantities of hoar frost.

Hibernation is one of the strangest phenomena of the animal world, and bears, especially the white bear of the polar regions, the black bear of North America, and the brown bear of Europe, agree in the curious habit of semi-hibernation. In the late fall of the season, the bears begin to eat heavily and soon become enormously fat, preparatory for the long winter of semi-sleep.

During the winter, at least for three months, the polar bear takes no food, but lives entirely upon the store of fat which her body had accumulated before she went into retirement. The same is true of many hibernating animals, but in case of the bears it is more remarkable because the mother bear must not only support herself but nourish her young for a long period without taking any food for herself.

Another good example of a ground-dweller is the aard vark of Southern Africa. He is as curious as his name, and scoops out immense quantities of earth to form his home. This dwelling might be termed a cave, as he heaps up the earth in the shape of a mammoth artificial ant-hill; on one side is the entrance, which is so skilfully formed that it looks far more like the work of man than of an animal.

His name is Dutch and means earth-hog. It is applied to him because his head looks somewhat like that of a pig. His claws are powerful and enormous, and with them he is able to dig into the hardest soil, and to destroy the giant ant-hills which are dotted over the plains of South Africa, and which can withstand the weight of a dozen men.

This strange creature sleeps during the day, and comes forth at evening to seek his food. The first thing he does is to burst a hole in the stony side of an ant-hill, to the utter dismay of its tiny inhabitants. As they run among the ruins of their fallen city, he throws out his slimy tongue and catches them by the hundreds. In a short time only the shell of a half-destroyed wall remains.

These once stately ant-homes metamorphosed into caves, form homes for the jackals and large serpents of the plains. The Kaffirs of Africa use them as vaults into which are thrown their dead. The aard vark outrivals, with his great claws, the most skilled burrowing tools of man. These animals are therefore rarely captured. It is not uncommon for a horse to fall into their excavations and be killed.

Miners, excavators, and underground dwellers teach us the great lesson that, while many of them sought the ground as a protection, and found there many difficulties to overcome, they not only have won in the great struggle of life but have so skilfully adapted themselves to their environment and surroundings as to become entire masters, even artists, in their methods of living.

VI

ANIMAL MATHEMATICIANS

"But what a thoughtless animal is man,--
How very active in his own trepan!"

--PRIOR.

Among the special senses of animals none seems more human than their knowledge of mathematics. A recognition of this quality in animals is encouraging because the new scientists are earnestly trying to build up a true knowledge of animal behaviour by studying them in the light of the new psychology. This will fill the place of the vast amount of misinformation which those skilled only in book-knowledge, without really knowing the ways of Nature, have builded. It will also record all the strange and curious facts about animals and their ways without insisting too much on rigid explanation. These new scientists are far different from their predecessors who tried to explain everything they did not understand about an animal's behaviour in terms of the scanty information gained by studying a few museum specimens. We might as well attempt to explain human nature from the study of an Egyptian mummy. The new method is simply to give the facts about an animal, and frankly admit that in many cases, such as are found in their knowledge of counting and numbers, we must leave complete explanation to the future when we shall have a greater fund of scientific data on which to base our conclusions.

It is an established fact that some animals can count, and that they have the faculty of close observation and keen discrimination. They learn to count quickly, but they do not fully appreciate the value of numerical rotation. Most of the arithmetical feats of trained animals are hoaxes regulated by their sense of smell, sight, touch and taste. But no one doubts their ability to count. I have known a monkey that could count to five. He played with a number of marbles, and I would ask for two marbles, one marble, four marbles, as the case might be, and he would quickly hand the number requested.

Another incident that will illustrate the point is the case of a mule owned by an old negro near Huntsville, Texas. The regular routine work of this mule was to cart two loads of wood to the town every day. One day the negro wished to make a third trip, but was unable to do so. When asked the reason, he replied, "Dat fool mule, Napoleon, done decided we had hauled enough wood fo' one day!"

Prantl claims that the time-sense is totally absent in animals, and that it belongs only to man, as one of the attributes of his mental superiority. However, many facts go to show that animals have not only a specific time-sense, but also a sense of personal identity which reaches back into the past.

Time-sense is very highly developed in dogs, cats, hogs, horses, goats, and sheep. They apparently are able to keep an accurate account of the days of the week and hours of the day and night, and even seem to know something of numerical succession and logical sequence. A friend in Texas had an old coloured servant, whose faithful dog had been trained to know that just at noon each day he was expected to carry lunch to his master. I have seen the dog on more than one occasion playing with children in the streets, suddenly break away without any one calling him, or any suggestion on our part as to the time, and rush for the kitchen just at the proper moment. No one could detain him from his duty. This same dog, however, would on Sundays continue to play at the noon hour. Surely, if any explanation is to be offered in such a case as this, it will imply as strict a sense of time as it does of duty.

A friend relates a case of a dog that went each evening to meet a train on which his master returned from the city. On one occasion the train was delayed two hours, and it was exceedingly cold, but the devoted companion remained until his master arrived. Innumerable instances of such all-absorbing affection, showing at the same time a sense of time, might be cited.

Dr. Brown gives a most remarkable example of a dog's ability to distinguish time. The story is of a female dog, though named Wylie, which was purchased by Dr. Brown when he was a young man, from an old shepherd who had long been in his employment. Wylie was brought to his father's, "and was at once taken," he says, "to all our hearts; and though she was often pensive, as if thinking of her master and her work on the hills, she made herself at home, and behaved in all respects like a lady.... Some months after we got her, there was a mystery about her; every Tuesday evening she disappeared; we tried to watch her, but in vain; she was always off by nine P. M., and was away all night, coming back next day wearied, and all over mud, as if she had travelled far. This went on for some months, and we could make nothing of it. Well, one day I was walking across the Grass-market, with Wylie at my heels, when two shepherds started, and looking at her, one said, 'That's her; that's the wonderful wise bitch that naebody kens.' I asked him what he meant, and he told me that for months past she had made her appearance by the first daylight at the 'buchts' or sheep-pens in the cattle-market, and worked incessantly, and to excellent purpose, in helping the shepherds to get their sheep and lambs in. The man said in a sort of transport, 'She's a perfect meeracle; flees about like a speerit, and never gangs wrang; wears, but never grups, and beats a' oor dowgs. She's a perfect meeracle, and as soople as a mawkin'.' She continued this work until she died."

Another most striking instance, showing animals' sense of time, is that related by Watson in which he tells of two friends, fathers of families, one living in London and the other at Guilford. For many years it was the custom of the London family to visit their friends in Guilford, always accompanied by their spaniel, Caesar. After some years a misunderstanding arose between the two families. The usual Christmas visits were discontinued; not, however, so far as the spaniel was concerned. His visits continued as before. On the eve of the first Christmas following the misunderstanding, the Guilford family were astonished to find at their door their London friend, Caesar. Naturally, they expected that he had come in advance of the family, and were happy in the thought of this unexpected reconciliation. All evening they awaited their friends, but none arrived. Nor did they the next day. Caesar had come of his own accord at the accustomed time, and remained with his friends for the usual number of days. This naturally led to a correspondence between the families, who thereupon resumed their former friendly relations. We do not believe, of course, that this dog counted the exact number of days to know when to start to Guilford, but he doubtless saw something to remind him of the past.

Sir John Lubbock once related before the British Association at Aberdeen how cards bearing the ten numerals were arranged before a dog, and the dog given a problem, such as to state the square root of nine, or of sixteen, or the sum of two numbers. He would then point at each card in succession, and the dog would bark when he came to the right one. The dog never made a mistake. If this was not evidence of a mentality at least approaching that of men, we do not know what to call it.

If there is any difference between an animal and a human mathematician, it depends upon special training. The animal never has the same opportunities to learn as the man. Many savages, for example, cannot count beyond three or four. Sir John Lubbock gives an anecdote of Mr. Galton, who compared the arithmetical knowledge of certain savages of South Africa and a dog. The comparison proved to the advantage of the dog.

There is no reason that a dog should not be taught arithmetic. And if one wishes to do so, it might be well to begin by making the dog distinguish one from two, allowing him to touch both once at the word one, and twice at the word two. Then he might pass on to six or seven. After he had progressed to ten, he might begin addition. At least the experiment would be interesting and conducive to learning the truth. Surely a knowledge of mathematics is no more wonderful than that of the ordinary pointer dog's ability to distinguish different kinds of birds. Certain of those wise dogs are trained to hunt only quail, while others hunt several varieties of game.

It should be remembered that all degrees of arithmetical aptitude are found in the human races, from the genius of a Newton and a Laplace to the absolute inability of certain of the Hottentots to count to three. These inequalities in the mathematical notions of different people should make us very cautious about saying that animals cannot count and have no sense of numbers. It is extremely probable that if we had a way of choosing those animals with a special gift for arithmetic, they would surprise us with their learning.

THE COYOTE CAN READILY DISTINGUISH WHETHER A HERD OF SHEEP IS GUARDED BY ONE OR MORE DOGS, AND WILL PLAN HIS ATTACK ACCORDINGLY.

THE ZEBU, THE SACKED BULL OF INDIA, IN SPITE OF ITS DOMESTICATION, HAS AN AGILE BODY AND A QUICK, ALERT MIND.

No one denies that animals are capable of distinguishing relative sizes and even quantities. They are not so skilled as the average human being in making these distinctions, yet when mentally compared to the state of Bushmen, Tasmanians, and Veddahs, who can count only two, and call it many, there is not such a vast gulf between them and mankind.

The zebu, or sacred bull of India, shows his mathematical qualities to a pronounced degree. When he grows attached to a small group of his kin, he will often refuse to leave them unless the entire group accompany him. When driven from his pen, if by chance one of his party is left behind he refuses to go--thus indicating that he is able to tell that the exact number is not with him. His affectionate and gentle disposition, not to mention his love of his offspring, would entitle him to rank among the most human of animals. No wonder he is worshipped in India, where the human side of animal life is understood and appreciated to a degree quite unknown to the Western world!

The fox and the wolf, and even the coyote, can readily distinguish whether a herd of sheep or cattle is guarded by three or four dogs, and whether there is one herdsman or two. They cannot tell the exact number of sheep, however; neither could a man without first counting them. Their knowledge of geometry is remarkable. They can orient themselves to the surrounding woods, measure distances, figure out the safest way of escape, and the power of the enemy even better than savage man. Yet in most of these problems, definite notions of number or figures have little part. A dog, when hunting, for example, on a prairie where he has to leap over ditches or quickly turn around a large tree, is able by a second's thought to do so without danger. He clears the wire fence, leaps the ditch, dashes through a closing gate, or escapes an infuriated enemy at a moment's notice. This natural wisdom is exercised spontaneously in him, it is the result of inborn theorems of which he may not even be aware, but which he uses with a sureness that defies the book-learning of all our teachers of mathematics. He uses speed, force, space, mass, and time with so small an effort, and by the quickest and shortest routes.

Suppose a wolf or a wild hog could not tell how many dogs were attacking it? There would be no way for it to defend itself. If four dogs attack it, they are counted and the tactics used that would be useless in other cases. If four dogs attack, two on each side, it retreats, with face toward the enemy. If a dozen dogs are in the attacking force, the hog becomes confused, loses all idea of number, and wildly bites at any enemy that comes nearest. Man in a similar condition would use practically the same tactics.

Cats undeniably count their kittens. If the mother loses one of three or four, she searches for it immediately. When dogs are chasing a hare, if they raise another, they become very confused, as if they did not know which to follow. Many shepherd dogs know if a sheep is missing from the flock and go to hunt it.

The efforts of scientific investigators, who work with so many learned theories, have been less successful in discovering the real facts about animals than of laymen, largely because the scientists have not yet learned that arithmetical notions are more difficult than geometrical ones. Our industrial civilisation has caused us to lose the idea of the insignificance that number has in animal life compared to the idea of size. Most animals have a remarkable sense of size; they measure time and distance better than civilised man. A hyena, for example, knows just how near he dare approach an unarmed man.

A sense of time is common among animals that daily eat at fixed hours. A donkey was accustomed to being fed at six o'clock in the morning, and when on one occasion his master did not appear on time, he deliberately kicked in the door to the barn and proceeded to feed himself.

Animals are capable of measuring lapses of time in which they are particularly interested. Houzeau claims that a female crocodile remains away from her eggs in the sand for twelve to twenty days, according to the species, but returns to the place exactly on the day they hatch.

Although we should hesitate to affirm that all animals have an extensive knowledge of figures and numbers, yet it can hardly be denied that the elephant, donkey, horse, dog, and cat, if given the proper training, become good mathematicians. It is undeniable that they have a love of mental acquisition, and it seems that the Creator has given to every animal, as a reward for its limitations in other respects, a definite innate knowledge and desire to advance educationally. There is in the breast of every animal an irresistible impulse which urges it to advance in the scale of knowledge. Where the animal is blessed with other mental powers, there is found a perfect harmony--of tact, intuition, insight, and genius--all that man himself possesses.

VII

THE LANGUAGE OF ANIMALS

"Who ever knew an honest brute
At law his neighbours prosecute,
Bring action for assault and battery
Or friends beguile with lies and flattery?"

The fact that all animals possess ideas, no matter how small those ideas may be, implies reason. That these ideas are transmitted from one animal to another, no one can doubt in the light of our present scientific knowledge. "Be not startled," says the distinguished animal authority, Dr. William T. Hornaday, "by the discovery that apes and monkeys have language; for their vocabulary is not half so varied and extensive as that of the barnyard fowls, whose language some of us know very well." The means by which ideas are transmitted from one animal to another can be rightly described by no other term than *language*.

It is evident that there are many kinds of language: the written; the spoken; the universal, which implies the motion, sign, and form language; the language of the eye, by which ideas are exchanged without words or gestures; and lastly, a mode of expression little known to the human world, but universal among animals. This language is spoken by no man, but is understood by every brute from the tiniest hare to the largest elephant; it is the language whereby spirit communicates with spirit, and by which it recognises in a moment what it would take an entire volume to narrate. In its nature it differs essentially from all other languages, yet we are justified in thinking of it as a language because its function is to transmit ideas from one animal to another. Every form of language is used by animals, and each has its own peculiar language or "dialect" common to its tribe only, though occasionally learned by others. All the emotions--fear, caution, joy, grief, gratitude, hope, despair--are disclosed by some form of language.

It would be interesting to know how the use of the word "dumb" ever became applied to animals, for in reality there are very few dumb animals. Doubtless the word was originally employed to express a larger idea than that of dumbness, and implied the lack of power in animals to communicate successfully with man by sound or language. The real trouble lies with man, who is unable to understand the language spoken or uttered by the animals.

The gesture language is commonly used by many of the tribes of Southern Africa, and some of the Bushmen are unable to converse freely after dark, because their visible gestures are needed as an aid to their spoken words. Only a few years ago there were almost as many different languages among the North American Indians as there were different tribes, and yet each tribe had a sign-language which any Indian in any part of the world might understand. In fact it was so simple that it might be practically mastered in a few hours, and through it one might converse with the Indians of the world without knowing a single word of their spoken language. And this is exactly what the animals do with their universal language.

Who does not understand the meaning of a dog when he approaches his master, after receiving a reprimand for some misdemeanor, with downcast head and lowered tail? Or who could fail to interpret the glee when he has done a noble deed and been praised by his master? His is the language of gesture and look, and is very similar to that in use by our deaf-and-dumb men throughout the world.

The Hindoos invariably talk to their elephants, and it is astonishing how they understand. Bayard Taylor says that "the Arabs govern their camels with a few cries, and my associates in the African deserts were always amused whenever I addressed a remark to the dromedary who was my property for two months; yet at the end of that time the beast evidently knew the meaning of a number of simple sentences. Some years ago, seeing the hippopotamus in Barnum's museum looking very stolid and dejected, I spoke to him in English, but he did not even open his eyes. Then I went to the opposite corner of the cage, and said in Arabic, 'I know you; come here to me.' I repeated the words, and thereupon he came to the corner where I was standing, pressed his huge, ungainly head against the bars of the cage, and looked in my face with a touch of delight while I stroked his muzzle. I have two or three times found a lion who recognised the same language, and the expression of his eyes, for an instant, seemed positively human."

Every one familiar with the habits of dogs believes that they have a language. Certain shepherds are quite particular about the company their dogs keep. This story is told of a couple of shepherds meeting in a market-place in Scotland, each accompanied by his dog, one of which was a sheep-murderer, the other a faithful and respectable dog. They seemed to strike up a great friendship, "and soon assumed so remarkable a demeanour in their conversation that their owners consulted together on their own account, and agreed to set a watch upon them. On that very evening both dogs started from their homes at the same hour, joined each other, and set off after the sheep." It is unquestionable that these dogs had a sufficiency of language to understand each other. The criminal had invited his innocent young friend to join him in his mischief, and they agreed upon the time to meet and each kept his appointment. It is likely that there was not an audible sound uttered during their conversation, but that they used the language of look and gesture, and while it was not understood by their masters, it was entirely comprehended by themselves.

Another instance of canine language is given by John Burroughs, who says that a certain tone in his dog's bark implies that he has found a snake.

There is an old maxim which says: "The empty wagon makes the most noise," and it is interesting to note that the loudest-mouthed and most loquacious of all the animals are the lemurs, who are the least intelligent members of their great family. They chatter, scream, squeak, and grunt from morning till night, and two of them can make more noise than a cageful of apes and monkeys. The orangs and chimpanzees, on the other hand, exceptionally wise and gifted linguists, seldom utter a word or cry, except under extraordinary circumstances, and then briefly.

Prof. Richard L. Garner, who has spent much time in studying the language of animals, has attracted a great amount of attention through his special study of the anthropoid apes. He has lived among these animals in a steel cage in their native haunts and has used a phonograph to record their language. Prof. Garner told recently of an exceptionally intelligent ape, named Susie, whose home used to be at the Zoological Park, under the care of the Zoological Society, and he claimed that Susie could speak "in her own language" at least five words. They were "yes," "no," "protest," "satisfaction" and "contempt."

Mr. George Gladden, writing in the *Outlook* on the chimpanzee's voice, did not exactly commit himself as to his belief regarding this matter, but he says: "Now, although Mr. Engeholm (for four years in charge of the Primates House in the New York Zoological Park) has not been able to discover that his apes use any language, correctly speaking, he is confident that the chimpanzees Susie, Dick, and Baldy comprehend the definite meaning of many words, and that their minds react promptly when these words are addressed to them in the form of commands. This capacity is more highly developed in Susie than in any other of the apes in this particular group....

"It is difficult, of course, to determine from the commands which an animal will obey precisely how many words employed in these commands are plainly understood; but I have endeavoured to do this tentatively in the case of Mr. Engeholm's commands to Susie, all of which I have seen her obey repeatedly and promptly."

Mr. Gladden enumerates about forty-three commands which he claims to have seen Susie obey promptly. And he further states that the belief which many students of animal psychology hold that an animal gets more of the meaning of a command from the gesture which accompanies the command than he does from the actual words by which he is commanded, is false, and he adds, "as to this, I can testify that of the forty-three commands ... thirty-six may be, and generally are, unaccompanied by any gesture whatever. How, then, does Susie comprehend those commands unless through her understanding of the meaning of the words in which they are conveyed?"

The distinguished phrenologist Gall had a dog whose memory was remarkable, and he thoroughly understood words and phrases. "On this subject I have made," says Gall, "the following observations: I have often spoken intentionally of things which might interest my dog, avoiding the mention of his name, and not letting any gesture escape me which would be likely to arouse his attention. He always exhibited pleasure or pain suitable to the occasion, and by his conduct afterwards showed that he understood perfectly well."

Col. W. Campbell in his *Indian Journal* gives two remarkable instances of language and unity of work among animals which he saw at Ranee Bennore, while he was on a hunting trip. He witnessed, one morning, a striking case of wolfish generalship, which in his belief proved that animals are endowed to a certain extent not only with reason but are able to communicate their ideas to others. He was scanning the horizon one morning to see if any game was in sight when he discovered a small herd of antelopes feeding in a nearby field. In another remote corner of the field, hidden from the antelopes, he saw six wolves sitting with their heads close together as though they were in deep conversation.

He knew at once that they were also seeking venison for breakfast and he determined to watch them. He concealed himself behind a clump of bushes, and the wolves who had evidently already decided upon their mode of attack began their manoeuvres: one remained stationary, while the other five crept to the edge of the field and one by one took the most advantageous positions, the fifth concealing himself in a deep furrow in the centre of the field.

The sixth, which had made no previous movements, dashed at the antelopes. The swift, graceful creatures, trusting in their incomparable speed, tossed their heads as if in disdain of so small an enemy and galloped away as though they were riding on the winds with their enemy far behind. But as soon as they reached the edge of the field, one of the hiding wolves sprang up and chased them in an opposite direction, while his fatigued accomplice lay down to recuperate. Again the light-heeled herd darted across the field, evidently hoping to escape on the opposite side, but here again they met another crafty wolf who chased them directly toward another of the pack. The chase had begun in earnest, the persecuted antelopes were driven from place to place, a fresh enemy springing up at every turn, till at last they became so terrorised with fear that they crowded together in the center of the field and began running around in diminishing circles.

During all this performance, the wolf which was hidden in a furrow in the centre of the field had not moved, although the antelopes had passed around and over him dozens of times. He well realised his time for action had not yet come and crouched closer and closer awaiting a signal from his fellow hunters to spring into their midst, and down one of the weakened antelopes.

At this point Col. Campbell shot one of the wolves, and the other five ran away and allowed the antelopes to escape. Surely no human combination could have shown greater reason and concerted action than was shown by the wolves under such conditions. Each had a particular post assigned, and evidently some means of communication was used in indicating their respective locations. Each had a definite part to play in the complex scheme--so that their language quite evidently expressed abstract ideas. That these ideas were carried out shows that the wolves were capable not only of laying ambitious plans for capturing prey, but of carrying them out as well.

"That beasts possess a language, which enables them to communicate their ideas," says Thomas Gentry, "has been clearly shown. It is just as apparent that they can act upon the ideas so conveyed. We have now to see whether they can convey their ideas to man, and so bridge over the gulf between the higher and the lower beings. Were there no means of communicating ideas between man and animals, domestication would be impossible. Every one who has possessed and cared for some favourite animal must have observed that they can do so. Their own language becomes, in many instances, intelligible to man. Just as a child that is unable to pronounce words, can express its meaning by intimation, so a dog can do the same by its different modes of barking. There is the bark of joy or welcome, when the animal sees its master, or anticipates a walk with him; the furious bark of anger, if the dog suspects that any one is likely to injure himself or master, and the bark of terror when the dog is suddenly frightened at something which he cannot understand. Supposing, now, that his master could not see the dog, but could only hear his bark, would he not know perfectly well the ideas which were passing through the animal's mind?"

There is no doubt that animals understand something of our human language. They may not be able to comprehend the exact words used, but it is evident they get the meaning to a certain extent. I once had a small Mexican dog sent me from Mexico; he seemed not to understand what was said to him, until a friend called who spoke to him in Spanish, whereupon he showed his delight and became at once a friend to the man who spoke his own language.

The Rev. J. G. Wood tells the following incident, which forcibly illustrates the ability possessed by animals to commune with each other. "While I was living in the country with a friend, a most interesting incident was observed in the history of the dog. My friend had several dogs, of which two had a special attachment to, and an understanding with, each other. The one was a Scotch terrier, gentle and ready to fraternise with all honest comers. The other was as large as a mastiff, and looked like a compound between the mastiff and the large rough stag-hound. He was fierce, and required some acquaintance before you knew what faithfulness and kindness lay beneath his rough and savage-looking exterior. The one was gay and lively, the other, stern and thoughtful.

"These two dogs were often observed to go to a certain point together, when the small one remained behind at a corner of a large field, while the mastiff took a round by the side of the field, which ran up-hill for nearly a mile, and led to a wood on the left. Game abounded in those districts and the object of the dogs' arrangement was soon seen. The terrier would start a hare, and chase it up the hill towards the large wood at the summit, where they arrived somewhat tired. At this point, the large dog, who was fresh and had rested after his walk, darted after the animal, which he usually captured. They then ate the hare between them and returned home. This course had been systematically carried on some time before it was fully understood."

Every animal has a definite language which is quite sufficient to express the desires and emotions of its nature, and to make them intelligible, not only to its own species, but also to other animals and sometimes to human beings. Those which do not actually speak by means of a voice, make signs or mimic understood things so as to be perfectly intelligible. If animals had no language, they could not instruct their young. The young of animals in a civilised country are far wiser than the old ones in wild, uninhabited countries. This can be explained only by the knowledge which the young receive from their parents.

It is not uncommon for animals belonging to widely different species to speak the same language, and thus become great friends. A friend in Texas once owned a cow whose sole companion was a small black goat. One day the young goat followed the cow home from her grazing place, and from that time on they were constant companions, even occupying the same stall in winter, sharing the same food, and always sleeping near each other.

If one shoots a monkey in South Africa, and wounds it, allowing it to escape, there usually come droves of its kinspeople, screaming and chattering the most diabolical language, seeking to revenge the wrong done their tribe. Nothing demonstrates plainer that they have a common language; otherwise, how could they understand that one of their number had been wounded? It is because of the communication of ideas by a common language among animals that hunters so fear to allow a wounded animal to escape at the beginning of their hunting season in certain localities. A wounded bear who escapes, for example, will spoil the entire season for hunters by spreading the alarm among his people.

American Museum of Natural History, New York

ROOSEVELT'S COLOBUS. THESE HORSE-TAILED MONKEYS CHATTER TOGETHER IN A LANGUAGE EXCLUSIVELY THEIR OWN, YET THEY SEEM TO HAVE NO DIFFICULTY IN MAKING THEMSELVES UNDERSTOOD BY OTHER MONKEY-TRIBES.

American Museum of Natural History, New York

A TAMED DEER OF TEXAS, WHOSE CONSTANT COMPANION AND PLAYMATE WAS A RABBIT DOG. BETWEEN THE TWO THERE DEVELOPED, NECESSARILY, A COMMON LANGUAGE.

Near our country home in Texas my sister found a very young red deer one morning just outside the garden, and bringing it into the yard, soon had a wonderful pet in this dainty spotted child of the woods. We knew that its mother was not far away, and so we placed salt and food just where the baby was found, to attract the mother's attention. In a few days, we saw the mother, and shortly afterwards five grown deer were seen eating the food we had placed for the mother. Evidently the news had been carried through the pine forests that it was safe for deer to come near our home. My sister's pet grew rapidly, and became a great friend of our yard dog. They often played by running races together, the deer would leap over the fence and the dog would chase him with great delight. Surely, they must have had a spoken common language!

No one claims that in the language of animals there are principles of construction such as we find in the human languages. The term Barbarian means those whose language is only a "bar-bar," and this is really all that the sound of an unknown tongue implied to the cultured Athenians. The neighing of horses, the howling of dogs and wolves, the mewing of cats, the bleating of sheep, the lowing of cows, the chattering of monkeys and baboons is nothing more nor less than their language. And it is quite as intelligible to us as is the chattering of the Hottentots of Africa. Because we do not speak the languages of our animal friends does not take away from the genuineness of the languages; we might as well claim that because our horse does not comprehend what we are saying, that we are not speaking a language!

Animals and men, under normal conditions, have been friends and companions since the beginning of time; and in order that they may convey ideas to each other, it is necessary for them to have some sort of means of communication.

As a matter of fact, animal language is quite often intelligible to man. Their language might be likened to that of a young child that cannot pronounce distinctly the words we commonly use; and yet we get the meaning from the intonation and gesture.

Any man who has ever owned a horse understands the meanings of his various actions and vocal expressions. There is the neigh of joy, upon returning home after a hard day's work, the neigh of distress, when he has strayed from his companions, the neigh of salutation that passes between two horses when they meet, and the neigh of terror when enemies are near. There is also the neigh of affection that is often given to his master when they first meet in the morning. Thus, spoken words are not necessary to express elemental feelings.

Elephants readily understand most of the words uttered by their masters. Menault tells of an elephant that was employed to pile up heavy logs. The manager, suspecting the keeper of stealing the grain set aside for the elephant, accused him of theft, which he denied most vehemently in the presence of the elephant. The result was remarkable. The animal suddenly laid hold of a large wrapper which the man wore round his waist, and tearing it open, let out some quarts of rice which the fellow had stowed away under the voluminous covering.

Animals have the power to make themselves understood by man, especially when they are in distress and wish man to help them. And they often combine to help one another. I was on a sheep ranch in western Texas once when one of the sheep came bleating up to the camp late in the afternoon. She uttered the most distressing calls. A friend, whom I was visiting, assured me that something unusual was wrong. Together we followed the sheep back to where she had been feeding in the pasture, she going forward in short spurts and continually looking back to see if we were coming. She finally led us to an old well, and we heard the plaintive voice of her young lamb that had fallen in. As the well had no water in it, and was only about six feet deep, we secured a ladder and in a few minutes the lamb was restored to its mother. She seemed delighted at the successful outcome of the accident. She had come and told us her troubles and got aid.

Cats are gifted linguists. By mewing they can just as plainly express a desire to have a door opened or closed as if they requested it in so many words. A friend has furnished me with an interesting account of her cat's ability to make herself understood. It seems that the cat, with her three small kittens, at one time slept in a box prepared for her in the kitchen. But one night when it was particularly cold, some one left the kitchen window open, and late in the night the cat went to her mistress's bed and mewed continuously until her mistress arose and went to the kitchen and closed the window. The cat was perfectly satisfied, as she had made her great need understood.

The ability that animals have to make their own language understood by man is not the only linguistic power they possess; as already mentioned, they are also capable of understanding something of human speech. There is no doubt that all domesticated animals understand the human language; the horse, dog, ox, and sheep comprehend a large part of what is said to them, though of course they may not understand the precise words used.

I once owned a rabbit dog, "Nimrod," and if he never understood another word of the English language, there is no doubt that he knew what the word "rabbit" meant. No matter in what manner or way I used the word, Nimrod was ready for a hunt, and yelped with glee at the thought of the chase that he was to have. I tested him over and over again by saying "rabbit hunt" gently; it thrilled him with delight, and while he was not very well educated in other things, he always lived up to his name.

The Rev. J. G. Wood speaks of the great individuality of character which he has observed in dogs, and that they unquestionably understand the human language. "There was in my pet greyhound 'Brenda,' there was in my dear lurcher 'Smoker,' and there is now in my dear lurcher 'Bar,' and in my three setters 'Chance,' 'Quail,' and 'Quince,' a refinement of feeling and sagacity infinitely beyond that existing in multitudes of the human race, whether inhabiting the deserts or the realms of civilisation.

"I cannot better define it than by saying that, if I give these dogs a hastily angered word in my room, though they have never been beaten, they will, with an expression of the most dejected sorrow, go into a corner behind some chair, sofa, or table, and lie there. Perhaps I may have been guilty of a hasty rebuke to them for jogging my table or elbow while I was writing, and then continued to write on. Some time after, not having seen my companions lying on the rug before the fire, I have remembered the circumstance, and, in a tone of voice to which they are used, I have said, 'There, you are forgiven.' In an instant the greyhound Brenda would fly into my lap, and cover me with kisses, her heart tumultuously beating. After she grew old, her joy at my return home after a long absence has at times nearly killed her; and when I was away, the bed she loved best was one of my old shooting-jackets, but never when I was at home."

The impassable gulf which the writers of old created between mankind and the animal kingdom was based mainly upon the belief that animals had no language, but this has been proved a mistake and no longer exists. In the light of modern knowledge and a better understanding of the marvellous theory of evolution, we are thoroughly convinced that there is no break whatever in the long chain of living beings. Man has no art, has developed no thing whatever, no mode of language or communication, that is not to be found in some degree among animals. They are capable of feeling the same emotions as human beings, and are therefore subject to the same general laws of life. No science has been more beneficial than psychology in proving that they are human in all ways; no discovery made by the human mind is so poetical and of such value as that which leads mankind to recognise some part of himself in every part of Nature, even in the language of animals.

This knowledge of all life is recognised by thinking men the world over, removing forever that artificial barrier by which, in his ignorance and prejudice, he has separated himself from his lower brothers, the animals, denying unto them even a means of intelligent communication. This recognition of the existence of a common language will go far toward establishing the universal brotherhood of all living creatures.

VIII

IN THEIR BOUDOIRS, HOSPITALS AND CHURCHES

"Never stoops the soaring vulture
On his quarry in the desert,
On the sick or wounded bison,
But another vulture, watching
From his high aerial look-out,
Sees the downward plunge and follows,
And a third pursues the second,
Coming from the invisible ether,
First a speck and then a vulture
Till the air is dark with pinions."

Many animals show a surprising knowledge of medical and sanitary laws, but these laws vary in the different species as much as they do among humans. Animals are divided into as many classes and social castes as are mankind; and those that have advanced beyond the nomadic life, and have fixed homes with servants and luxuries, naturally are more refined in the matter of their personal care.

Science may yet prove that the old legend of the mermaid sitting on a rock, with a glass and comb in her hand, was not so far from truth as we imagine. No doubt, the bright-eyed seals looked like sea-maidens to many ancient mariners. The originator of the mermaid stories had possibly seen seals making their toilettes. These beautiful and affectionate human-like creatures of the water, wear, attached to their front flipper, a handsome comb-like protuberance. When they rest on the rocks, they use this little comb to brush the fur on their faces; and the Northern fur-seals, when the weather is warm, use their flippers as fans. The secret of teaching seals to play tambourines is due to their desire to comb their fur and fan themselves!

Members of the cat family are, perhaps, the cleanest of all animals, with the exception of some of the opossums. Lions, panthers, and pumas dress themselves very much as the domestic cat performs her toilette. They use their feet, dipped in water, as wash cloths, and their tongues as combs and brushes. Hares also use their feet to wash their faces, and this they do very often, to keep their exquisite hair in perfect condition. Dogs enjoy wiping their coats against green grass and shrubs.

Certain animals are so fastidious that they have community beauty-parlours! Goats, deer, giraffes, and antelopes, for example, are very particular about their personal neatness and cleanliness, and they come together to assist each other in making toilettes. One of the reasons that animals suffer so much in captivity, especially when alone, is that they have no one to help them dress, and some of them, such as the giraffe, cannot reach all parts of their bodies. I have seen a young guinea pig that had been rescued from a mud puddle being cleaned by both of his parents. Water-loving animals, like the beavers, seemingly take great pride in their toilettes, and in this respect they show more human traits than any other animal.

It is a general belief that animals are quite care-free, and that when they awake in the morning there is nothing for them to do but play or wander about. This is a mistaken belief, for they have to dress themselves, and this not only means a bath in many cases, but a smoothing out of their fur and hair. Some are shy and seek the darkest places to dress themselves, others, like the dog and cat, seek the hearth. Every one has possibly seen a cow and horse licking each other, and it is generally believed that this implies special friendship between the two, but this idea is incorrect; it only implies mutual aid in making their toilettes. They have a beauty parlour, and thus aid each other. In no way are animals better prepared to teach man than in their methods of personal cleanliness, and this means health. Their utilisation of clay, dust, mud, water, and even sunshine to keep their health, far exceeds that of mankind. In fact, man's first knowledge of simple, natural health remedies came from animals. This wisdom they have acquired by ages of instinct and reason, for theirs has been the normal life, whereas man's is often abnormal. Each animal is his own specialist. However, when an animal becomes too ill to doctor himself, he is treated by another. I have seen a horse licking the wound of one of his fellows to stop the pain.

American Museum of Natural History, New York

WATER-LOVING ANIMALS, LIKE THE BEAVERS, SEEMINGLY TAKE GREAT PRIDE IN THEIR TOILETTES. THEIR FUR IS ALWAYS SLEEK AND CLEAN.

American Museum of Natural History, New York

GREAT FOREST PIGS OF CENTRAL AFRICA. LIKE THE COMMON DOMESTICATED HOGS, THEY WILL SEEK A CLAY BATH TO HEAL THEIR WOUNDS.

Animals know better than man what kind of food they need, for the simple reason that their tastes are natural, while man has allowed his to become perverted. In times of sickness absurd practices have been observed. Ice-cream and buttermilk, for example, were for ages refused to typhoid fever patients, while to-day they are generally used under such circumstances. But the natural desire for sour and cold things was always in evidence; animals have always depended upon these desires.

Among them are skilled dietitians, who restrict their diet in case of illness, keep quiet, avoid all excitement, seek restful places where there is plenty of fresh air and clean water. If a dog loses his appetite, he eats "dog grass," while a sick cat delights in catnip. Deer, goats, cows, and sheep, when sick seek various medicinal herbs. When deer or cattle have rheumatism, they invariably seek a health resort where they may bathe in a sulphur spring and drink of the healing mineral waters. They also know the full value of lying in the warm sun.

Cats are skilled physicians, and have various home remedies, such as dipping a feverish foot into cold water, or lying before a warm fire, if they have a cold. Many animals know how to treat a sore eye--by lying in the dark, and repeatedly licking their paws and placing them over the afflicted member.

How wonderful would the human race become, if it had the strength of a lion, the power of a bear, the wisdom of an elephant, the cleverness of a fox, and the health of the wild boar! But these qualities are found chiefly among the animals because of the marvellous knowledge of the laws of health and self-preservation.

John Wesley claimed, in his directions on the art of keeping well, that many of the medicines which were used among the common people of his time were first discovered by watching animals in their medical practices to cure their ills and pains. "If they heal animals, they will also heal men," he claimed. The American Indians learned most of their cures from watching animals, especially the cure of such diseases as fever, rheumatism, dysentery, and snake-bites. A rheumatic old wolf would bathe in the warm waters of a sulphur spring; a sick and feverish deer would eat the fresh leaves of healing ferns, while a wounded hog or bear would always seek a red-clay bath to heal the wounds. Sick dogs will invariably eat certain weeds, and an unwell cat will seek healing mints and grasses.

Old hunters tell us that a deer after having been chased for several hours by dogs, and after having escaped them by swimming a cold stream, will, upon reaching safety, lie down in the ice and snow. If a man did such a thing, he would immediately die. But not so with the deer, for he will arise about every hour and move around to exercise himself, and on the morrow he is perfectly well. The same animal, shut up in a warm barn for the night, as has many times been demonstrated with circus animals, will be dead by morning.

From this natural method of healing, mankind may learn much, and especially as it pertains to the treatment of extreme heat, cold, exhaustion, and paralysis of the muscles, and most especially sores and wounds. I have seen a wounded hog that had been badly bitten by a dog, wallow in rich red mud to stop the flow of blood.

It is a common practice for a raccoon actually to amputate a diseased leg, or one that has been wounded by a gunshot, and wash the stub in cool flowing water. When it is healing, he licks it with his tongue to massage it, and also to stop the pain and reduce the swelling. This wisdom is often classed by the unknowing under the term instinct, whereas it displays no less skill and knowledge than that of our modern surgery. The intelligence of the raccoon stands very high in the animal world.

Foxes, when caught in a trap, will very often gnaw off a limb. This requires a special power and a moral energy that few men possess.

William J. Long, in the *Outlook*, tells of an unusual proof of animal surgery in the case of an old muskrat that had cut off both of his forelegs, probably at different times, and had grown very wise in avoiding man-made traps, and when found, had covered the wound with a sticky vegetable gum from a pine tree. "An old Indian who lives and hunts on Vancouver Island told me recently," said Mr. Long, "that he had several times caught beaver that had previously cut their legs off to escape from traps, and that two of them had covered the wounds thickly with gum, as the muskrat had done. Last spring the same Indian caught a bear in a deadfall. On the animal's side was a long rip from some other bear's claw, and the wound had been smeared thickly with soft spruce resin. This last experience corresponds closely with one of my own. I shot a bear years ago in northern New Brunswick that had received a gunshot wound, which had raked him badly and then penetrated the leg. He had plugged the wound carefully with clay, evidently to stop the bleeding, and then had covered the broken skin with sticky mud from the river's brink, to keep the flies away from the wound and give it a chance to heal undisturbed. It is noteworthy here that the bear uses either gum or clay indifferently, while the beaver and muskrat seem to know enough to avoid the clay, which would be quickly washed off in the water."

Animals not only know how to doctor themselves when they are sick, but some of them, such as the fox, have learned how to make artificial heat by covering green leaves with dirt. And while they do not make fire, their homes are often heated in this practical way, and thus sickness avoided. Domestic horses and dogs wear hats in summer, and possibly in the future they will learn the enormous importance of wearing clothes! Trained monkeys already take great delight in dressing up, and dogs like smart suits.

Monkeys show the greatest interest and brotherly love when one of their number is injured. Watson tells of a female monkey that was shot and carried into a tent. Several of her tribe advanced with frightful gestures, and only stopped when met with a gun. The chief of the tribe then came forward, chattering and remonstrating vigorously. But as he came nearer, there was every evidence of grief and supplication for the body. As he was given the body, he affectionately took it in his arms and slowly moved to his companions, and like a silent funeral procession they all walked away.

Nor does their interest cease with life, for we are told by no less authority than Col. Theodore Roosevelt of a large grizzly bear that was discovered lying across the trail in the woods. The hunter shot her as she was preparing to charge him, and later he examined the spot where she was lying, and found that it was the newly made grave of her cub. Evidently some animal had killed the cub in her absence, and she, in her grief, was determined to avenge the wrong by lying in wait for the enemy.

Public meetings for civic council and religious worship are not confined to man alone. In Macgrave's *History of Brazil* we are told of a species of South American monkey known as the ouraines, which the natives call preachers of the woods. These highly intelligent creatures assemble every morning and evening, when the leader takes a place apart from the rest and addresses them from his pulpit or platform, Having taken his position, he signals to the others to be seated, after which he speaks to them in a language loud and rapid, with the gestures of a Billy Sunday, the audience listening in profound silence. He then signals again with his paws, when all cry out together in apparently confused noises, until another signal for silence comes from their leader. Then follows another discourse, at the close of which the assembly disperses. Macgrave attempts no explanation as to the object of these addresses; but if his accounts be true, surely they must have as much meaning for the monkeys as many of our public lectures and church services have for us! No doubt much of the advice imparted concerns the personal and collective welfare of the tribe members.

IX

SELF-DEFENCE AND HOME-GOVERNMENT

"In the days of yore, when the world was young,
Sages of asses spoke, and poets sung;
In God's own book we find their humble name,
Some enrolled upon the scroll of fame."

There is no phase of animal life which is more interesting than that through which Nature governs and protects her children. Each and every species of animal possesses the method of self-defence and protection best adapted to it. Most of the larger animals are of themselves so powerful that they need no protection other than that afforded by their strength, while most of the weaker and less aggressive animals are provided with some special method of defence.

The tiger, lion, panther, and wolf have formidable claws and teeth; while the shark has such immense jaws that he can sever the head of a goat at one bite. And most of them are in reality tyrants. They rule by tyranny--the oppression of the weak by the strong, whether that strength be physical or mental,--a trait as common in animals as in man. Among the animals it takes the commonest form, and they not only oppress the weak, but actually kill and eat them, even though they oftentimes are members of the same family. They are exactly like human cannibals, no better and no worse.

Flight is perhaps the simplest and most natural method of defence. The swifter animals, however, such as deer, gazelles, and hares, which may easily escape by running their fastest, do not always use this method, but have other means so ingenious as to be real arts. Wolves, when they see that they are outnumbered, will sometimes escape by following the exact tracks of a single leader through the snow, and from all appearances only one has passed the way over which a hundred may have gone. Hares will separate and run in opposite directions, while gazelles, if too closely pursued, will jump to one side and lie flat on the earth to escape notice, and as soon as the enemies have passed, run in the opposite direction.

It oftentimes happens that aggressively disposed animals, like cowardly men, are apt to try battle with the unlikeliest adversaries. A missionary from India tells the story of an alligator who was enjoying a noonday sleep on the bank of a river, when an immense tiger emerged from the jungle, made straight for the sleeping saurian until within leaping distance, when he sprang on the alligator's back, and gained a strangle hold before the sleeping monster could awake. At first the tiger was master, for the alligator could not bring his huge jaws into action, and while lashing viciously at the tiger with his tail, he was dragged into the jungle. What happened there no one could see, but in a few moments the tiger dashed out of the jungle and disappeared in the cane brakes, and the alligator reappeared and crawled into the water.

The ape and the baboon are the most skilled of all animals in making their flight. They use every method known to man, and because of their swiftness of action excel man in certain ways. Like man, in the face of danger, they show great bravery and never lose their presence of mind. The ape is fast disappearing before man, but against other animals and Nature he can well protect himself. He is even braver than the lion, who in captivity allows himself to be petted, but rarely is this true of the ape, and then only when conditions seem insurmountable.

In making his escape from an enemy, the ape directs his flight in the most self-possessed and human-like way, never losing his head, and taking advantage of the first shelter or protection that he meets; if the young, or females, or aged linger behind, a strong army of males bravely returns to rescue them at the danger of losing their own lives. Many of their brave deeds, if recorded in history, would compare favourably with those of mankind! Too often has a poor, sickly ape, which by his very feebleness allowed himself to be captured and placed in a zoo, been compared to human beings. Even in spirit and movements he has been considered as a human caricature and heaped with ridicule. We have continually considered his defects, without noticing his better qualities. We would have a much higher idea of his great family, if we would take a human derelict and compare him to an ape ruler! This comparison would be more just.

Certain of the baboon tribes which live among the rocks of high mountains and cliffs, if pursued by enemies, protect themselves by ingeniously rolling immense stones down upon their foes. They also hurl with great force small stones about the size of one's hand. As these tribes have each from one hundred to three hundred members, they constitute a formidable grenade army!

In addition to their skilled methods of flight, the baboons, apes, and monkeys come next to certain of the cat tribes as the greatest fighters in the animal world. This is astonishing when we remember that these animals are not professional warriors, nor do they have to fight to obtain their food. Their greatest defence is their quickness and powers of biting. When they are attacked by a dog, they usually bite off a foot or an ear, or leave him minus a tail!

One of the bravest and fiercest of fighters is the bull-dog. Three of these animals together have been known to capture and hold a large bull. Deer, when fighting among themselves, often play more than anything, and are not serious. Red deer seldom injure one another with their long antlers, but they could easily kill a dog or even a man. Stags, however, often fight to death, in some instances locking horns and tumbling over a precipice.

The most ingenious of all the horned fighters is the sable antelope, whose clever system of self-defence might well be taught in war-schools. His horns are long, sharp-pointed, and bend backwards. When wounded, or attacked by wolves or dogs, he lies down, and scientifically covers his back by rapid fencing with his pointed horns. He can quickly kill any dog that attacks him in this way.

Occasionally great battles take place between a buffalo and a lion, or more often two or three lions attack a buffalo, who rarely escapes them. The strength of a lion is almost beyond our comprehension when we remember that one can actually carry a cow over an ordinary-sized fence.

American Museum of Natural History, New York

THE ROCKY MOUNTAIN GOAT HAS MANY MEANS OF DEFENCE, NOT THE LEAST OF WHICH IS HIS AGILITY IN CLIMBING TO INACCESSIBLE PLACES.

American Museum of Natural History, New York

WILD BOARS ARE AMONG THE MOST FEROCIOUS OF ANIMALS. BY MEANS OF THEIR GREAT STRENGTH ALONE THEY ARE WELL ABLE TO DEFEND THEMSELVES.

A most unique fighter is the giraffe. He has neither claws nor sharp teeth with which to defend himself; so, if he gets angry with one of his kind, he deliberately uses his long neck like a pile driver would use a sledge hammer. Swinging it round and round, he lets his head descend upon his adversary like a heavy ax! The two animals use the same kind of tactics, and bracing themselves so as to stand the blows, they fight until one has to give in. Their heads are furnished with two small knob-like horns which only protect them from the heavy blows without serving as offensive weapons.

Most singular and amusing of all methods of self-defence are those which entirely depend for their efficiency upon bluff, or pretence. The chameleon, for example, erects his snake-like hood, though he is harmless, and at the most could scarcely injure the smallest animal. Equally curious are the methods of skunks and polecats, which project against enemies a highly disagreeable fluid.

Passive modes of defence are as many and varied as are the active; one of the strangest and most inexplicable of these is that known as spontaneous amputation, technically termed autotomy. The lizard, for example, when captured, will abruptly break loose his tail in order to escape; and certain wood rats, when caught, loosen the skin on their tails and deliberately slip away. Autotomy not only permits flight, but also defends the animal against the most adverse conditions. Nearest akin to this--defence by means of amputation--is the practice of bears and raccoons of amputating their limbs when caught in steel traps.

Mimicry, which is treated under another chapter, comes under the head of passive defence, and form and colour play an important part in it. Strangely enough, animals which have never resorted to mimicry as a means of protection, when associated with others who practice it, take on the habit themselves. This may possibly be due to the fact that new enemies are constantly arising.

As human sharpshooters dress in garments of the same colour as the woods in which they hunt, so many animals use this principle of imitation. The colour of most animals is very similar to their surroundings. This enables them to lie in wait for prey, a practice as old as the hillsides with animals. They have learned the extreme value of silence, and that they must remain at times motionless. This is especially noticeable with crocodiles, which wait for whole days without moving, concealed in the water or deep grass, until their prey comes within striking distance, when they pounce upon it. The same is true of the python snake, which hangs from a tree so immovable that he appears like a vine or a branch of the tree. If an animal attempts to pass, he drops upon it.

Perhaps the most unique and successful method of passive defence is the feigning of death, or "playing 'possum" met with in several animals, such as the red fox, the opossum, occasionally the elephant, and several of the snakes. On many occasions I have been 'possum hunting in the South and found my dog barking at an apparently dead 'possum. As soon as these animals are approached by larger and stronger enemies, they drop absolutely motionless on the ground and close their eyes as though they were dead. Here they remain until the enemy either destroys them, carries them away, or leaves them alone. If left alone for a few moments, they immediately spring to their feet and make their escape.

Elephants often feign death when captured, in order to gain their liberty. Animal catchers tell many interesting tales of elephants feigning weakness from which they fall to the earth and later apparently die. In many instances the fastenings are removed from their legs and head and the carcass is abandoned as useless, when to the utter astonishment of all--before the captors get out of sight--the animal springs up and dashes away to the forest, screaming with joy at the triumph of its deception.

Many animals deliberately assume a frightful, terrifying or grotesque appearance. This they do by inflating their bodies, by erecting hair, skin, or folds, or by unusual poses. Darwin speaks of the hissing of certain snakes, the rattle of the rattle-snake, the grating of the scales of the echis, each of which serves to frighten or terrify the enemy.

Bluffing is another form of defence that many animals use. The cobra, for example, when disturbed, raises its immense hood in a most terrifying attitude! Many of the lizards use the same tactics; while the horned toads of America when disturbed actually eject blood from their eyes. Every one is familiar with the cat's habit of raising the fur on his back when molested by a dog. All bluffing animals, when in danger, try to assume a pose that will make them look most dangerous and impressive to their enemies, and there is little doubt that in most cases they succeed very well, for we have all seen a dog slink away from a menacing cat.

The elk or moose, whose home is in the northern part of America and Europe, is a powerful and large animal, sometimes seven feet in height, and is able to endure much cold. He has many enemies among animals and mankind, and during the summer season he is quite able to protect himself, but in winter there is considerable danger from hordes of wolves. This is especially true just after a heavy snowstorm, if the snow is wet and melting. When it is dry and frozen, he can travel over it with great speed, and this he does by a most unusual trot which carries him along much faster than the trotting gait of a horse. Thus he is able to escape the hungry, carnivorous wolves, whose courage increases with appetite. If crowded too close, he is able also to protect himself by the most terrific blows of his fore-feet.

But when the spring weather sets in, and the snows begin to melt underneath, leaving the upper crust sufficiently strong to support the weight of lighter and smaller animals, such as wolves, especially when they travel swiftly, he is in great danger. For with every step he sinks to the belly in the snow, while his enemies can walk right up to his head and shoulders without his being able to strike or paw them with his dangerous hoofs. The advantage seems to be with the wolves, and if ever they bring the moose to bay in the snow, his life is doomed. For they care little for his arrow-like horns, but boldly jump at his throat and kill him. Herein comes the elk's wisdom--he deliberately sets to work, before the snow melts, and builds for himself and family an elk-yard, which is nothing more than a large space of ground on which the snow is smoothed or trampled down until it becomes a hard surface on which he can walk; it is also surrounded by a high wall of snow, through which are certain exits that allow him to pass out, if he desires. All the enclosed space is not smoothed down, but parts of it only are cut up into roads through which he may pass very swiftly. Woe unto the daring wolves that enter his snowy fortification--his "No Man's Land"--- for sure death awaits them!

A sense of law, order, government; the sacredness of family ties--all these aid in the protection of animals. Family life with them originated just as it did in the human world. The social instinct and the moral sentiments which arise from social relations in man and animal are the same. Moral obligations, especially in relation to family ties and conjugal unions of animals, are in many cases sacred binders to such ties. The bear, for example, is proverbial for his conjugal faithfulness. The married life of most animals is strictly moral, and most of them are monogamists and have reached the highest form of family association and life.

In those places where they live promiscuously, it gives them the same protection in herds as it does among our lower savages. Cattle, sheep, and horses unite for mutual protection; wolves band together in packs; and after they have been domesticated there is still not only a strong desire to band together for social purposes, but also to hold courts of justice. It sometimes happens that an angered husband takes the law in his hands, like uncivilised men, and beats his wife.

In the development and organisation of social and civil life the horse and the goat hold the foremost position. It corresponds to that of man among the lower animals. They do not believe in monarchies, but strictly in republics, or rather, a democracy where all power comes from the working class. The claims of the working class to the exercise of supreme control in all political affairs are practically realised. Among a herd of wild Arabian horses, the leading stallion, or so-called king, is really only the father of the tribe; his functions are paternal rather than regal. If he may be said to reign in a certain sense, the true workers rule, and his scouts and sentinels obey his wishes which the workers have influenced and formulated.

The existence of but one king leaves no room for dynastic troubles and rivalries which disturb, so often, our human countries and empires with such dreadful results. If two rival kings arise at the same time in a herd of horses, instead of forming factions in the state which end in civil war, they fight it out personally until one of them is killed or defeated. Once in a great while the other horses intervene, and drive the less desirable, or the false-claimant of power, away from the herd and its grazing territory. In these troubles the real king has little or no power, all activities are carried on by the workers.

If by chance he dies or is captured, another king, chosen by the herd, immediately assumes the kingship. It is a well-known fact that if the king of a herd of wild horses is caught, it is not uncommon for his herd to remain as near him as possible, and in their attempt to release him are often trapped themselves. The king has no heirs, either apparent or presumptive, and no right of succession is recognised. Any member of the herd, provided the workers choose him, may become the king, as every American school boy is a possible president of the United States.

Among many animals there is a perfect social and industrial organisation in which the division of labour is far better adjusted than in many human organisations. This, of course, is the result of gradual growth and evolution just as it is in the human species. This can easily be proved among animals by their more primitive and savage habits. Monkeys, for example, in civilised monkey communities, differ very greatly from those of wilder and less trained districts. They are constantly changing their habits, becoming more and more civilised by improving their methods of work and their moral and religious life as well. In many cases they have ceased to kill members of their own tribe for small offences for which they used to kill, and the cleanness and beauty of their home lives seem to increase with the years.

It oftentimes happens, however, that powerful ape and baboon colonies relapse into barbarism, and roam, plunder, rob and murder, like a pack of uncivilised wolves or hyenas. They seem all at once to forget their peaceful industries and lose all desire for clean and right living. And strangely enough, when they once turn bad, they seldom reform. Some naturalists believe that they are led astray by a wicked king or ruler who comes into power; the natives believe the evil spirits have suddenly taken possession of them.

There is unquestionably, in the life of many tribal animals, a definite historical connection between the mother tribe and its colonies. This relation extends to the tribes of tribes, and thus there is an international relationship between the various members of a large number of tribes. These communities share the same likes, dislikes, hatreds, and aspirations. A missionary friend told of his experience with monkey folk, and how once, when hunting, his gun was accidentally discharged, instantly wounding a large semi-tame baboon near his home. He hastened to help the injured animal, but saw that the relatives had crowded around and were terrorised, as they thought it was intentional. They not only followed him to his home, but returned in the night and actually tore his fence down. For months he was afraid to leave his wife alone during the day. And the natives reported that large tribes of monkey folk immediately came into the community from remoter regions and were distinctly on the war path. It was evident that their unjust antipathy was extended to all the kinspeople.

This is evidence of hereditary enmity, such as is common among families, tribes, and clans, and it often takes the form of feuds, which are still in vogue in the mountainous counties of the South. The baboons had suffered wrongs and never forgot it, and it was transmitted to their offspring.

American Museum of Natural History, New York

BRONTOSAURUS. THE ANIMALS THAT SEEMED BEST EQUIPPED TO DEFEND THEMSELVES ARE THE ONES THAT, THOUSANDS OF YEARS AGO, BECAME EXTINCT.

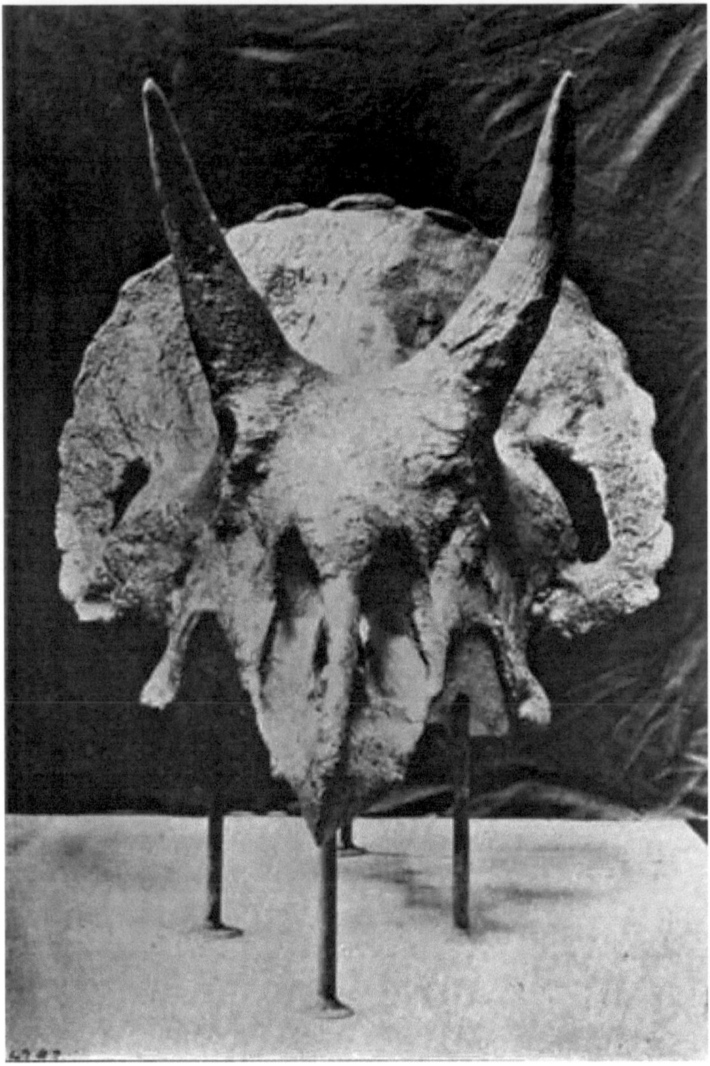

THIS PREHISTORIC MONSTER WAS EQUIPPED NOT ONLY WITH A PAIR OF STRONG HORNS, BUT WITH A SHIELD BACK OF THEM AS WELL.

The ability to use weapons, tools, and war instruments is not exclusively human. Even fish are capable of reaching their prey at a long distance. The *toxotes jaculator*, which lives in the rivers of India, and feeds upon insects, cannot afford to wait until the insects which thrive upon the leaves of aquatic plants fall into the water. So as he cannot leap high enough to catch them, he fills his mouth with water and squirts it at an insect with such aim and force that he rarely fails to knock the insect into the water where he can easily catch it. Many other animals squirt various liquids, occasionally in attack, but most times in defence. The fish makes a veritable squirt-gun of his mouth.

Beavers use sticks, chips, and even stones in building their dams; and their engineering abilities are astounding. They are also capable of meeting emergencies, as shown by the following incident. A farmer in Michigan discovered one morning, just after a flood, that all his potato sacks, which had been hung on a back fence to dry, had suddenly disappeared. A few days later he found them in a nearby beavers' colony, used in rebuilding their dam, which had suddenly overflowed. The beavers wasted no time, when they discovered their danger, in meeting the emergency by using the sacks to prevent the destruction of their home.

Monkeys make skilled use of clubs and stones in capturing their prey and fighting their enemies.

The skill with which some of them throw pebbles would lead us to believe they have already reached the degree of civilisation that many tribes of savages had reached only a few years ago, when they learned to use the boomerang and lasso. Some naturalists claim that monkeys actually set pitfalls for their enemies and lie in wait for them to be caught, just as a hunter would do.

Elephants also know the value of clubs in warfare, and will often use a broken limb of a dead tree as a weapon of defence. The story is told and vouched for by Mr. William B. Smith that on his farm, near Mount Lookout, a few years ago a donkey grazed in the same pasture with a ferocious bull. He was frequently attacked by the bull, and always got the worst of the fight. His feet were no match for the bull's horns, but one day the mule grabbed a long pole in his mouth, and, whirling it about, almost killed the bull, and henceforth the two lived on the best of terms in the same pasture.

I have a friend who owns a cow that knows exactly how to lift an iron latch to the barn door with her tongue and open the door. Innumerable times she has opened a gate in the same way to permit her calf to go free with her. So skilled is she in the manipulation of doors and latches that we are tempted to believe in some previous state of existence she was a professional lock-picker!

Cats and dogs are famed for their ability to open doors by pulling latch-strings. And not a few cats show a strong desire to study music by walking up and down the keyboard of a piano!

Monkeys who live near the seashore show wonderful aptness in opening oysters and shell-fish with sharp stones, exactly as a man would do. Monkeys have already reached the degree of civilization where they select the stones best suited for their work, and from their progress in the past it is reasonable to believe that in the near future they will not only be able to make their own tools--thus placing themselves on a mental footing with our flint-chipping ancestors of the early stone age,--but will also learn the use of fire and eventually the use of guns and ammunition, which marks one of the most important epochs in the evolution of the human species.

The chimpanzees, gorillas, and apes of the African forests have many times been observed in the act of piling brushwood upon the fires left by travellers, and though they do not know how to kindle a fire, they have learned how to keep it burning. The tame ones soon learn how to ignite matches, and often do great harm by starting forest fires.

But they show quite as much intelligence about the use of fire as the average small child. In fact, it has been thought by a number of great scholars that man had not yet made his appearance upon the earth in the miocene age, and that all the marvellous chipped flints of that age belong to semi-human pithecoid apes of wonderful intelligence. There is surely nothing in the facts of natural history, nor in Darwin's theory of evolution, that makes such a supposition unbelievable.

Baboons use poles as levers, stones as hammers, and seem to understand the more simple mechanical devices. Prantl claims that man is the only animal capable of using fire but not a few baboons know how to strike a match, heap dried leaves over the blaze to make it burn, and then heap on dead wood to feed the fire. This knowledge with them, exactly as with primitive peoples, is a product of long experience and does not show any mathematical truths or principles any more than making a direct cut across a field implies "knowledge of the relation of a hypothenuse to the two other sides of a right-angled triangle." This is what Prantl calls "spontaneous mathematical thinking."

I knew of a tame ape in Chicago that learned to swing from the end of a clothes-line and seemed to enjoy it very much. The line was just the right length and properly hung so as to allow the ape to swing out from a kitchen window and touch the ground. Just for fun, some one cut a piece from the line so that he could not reach the ground; immediately the ape hunted another piece of cord, tying it to the end of his line so as to increase its length, and much to his delight, continued to swing on the line.

The distinctive features of animal protection and home government, especially in the higher groups, may compare favourably with any of the methods used by civilised man. This is true both of their offensive and defensive contrivances and for their monarchies and republics. They use shells, scales, plates of every kind, with innumerable modifications for various purposes--spines and allied armaments--all shapes and sizes; poisonous secretions, deadly odours, strong claws and teeth wielded by strong muscles, and form colonies that are more than a gregarious association. In most cases, they have communities composed of individuals living individual lives, yet which act in cases of need as one unit.

X

ANIMAL ARCHITECTS, ENGINEERS, AND HOUSE BUILDERS

"The heart is hard that is not pleased
With sight of animals enjoying life,
Nor feels their happiness augment his own."

The most popular and perhaps the most interesting department of natural-history study is that which treats of the manner in which animals utilise the various materials of the universe for purposes of protection, for war and defence, for raiment, food, and even the luxuries of life. Man, by his superior power of adaptation, excels the lower animals in providing for the comforts of life; but, on the other hand, in such practical arts as engineering and domestic architecture man frequently finds himself an amateur in comparison. With all man's inventions he has not been able to equal some of the remarkable results produced by some animals. The beaver, for example, shows a more profound knowledge of hydraulics than man himself. The power possessed by these craftsmen, not only in felling trees, but in duly selecting the best places for making homes and in appropriating substances suitable for their needs, is a never-ending marvel!

Nowhere can we find a greater animal-workman than the beaver. He belongs to the great burrowing family, and is also extremely graceful in the water. Long ago he learned the advantages of co-operation, and he unites with his fellows in building dams of felled trees, which have been cut up into suitable length for use in damming up water places. These are skilfully placed, and with the aid of mud, control the level of the water in selected places as efficiently as man could do. As a social animal, the beaver should be ranked among the first; of course, the various marmots are extremely sociable, but they ordinarily live quite independently of each other, except in cases where they chance to congregate because of favourable conditions. The beavers, on the other hand, thoroughly understand the benefits of united labour, and work together for the good of the community.

Beavers, if their skill were generally known, would have a great reputation among their human friends. Recently, at the New York Zoological Gardens, a visitor was pointing out different animals to his little son, and when he came to the beaver pond, referred to two of these dam-builders and tree-cutters, which were swimming through the water with large sticks in their mouths, as big rats!

Young beavers make their appearance in May, and there are usually from four to eight to a family. These kittens, as they are called, are odd looking little fellows, with big heads, large sharp teeth, flat tails, like little fat paddles, and delicate, soft, mouse-like fur, not at all coarse like that of their parents. If taken at an early age they make nice pets and are easily domesticated. In the early days of American history it was not uncommon to see one running around an Indian lodge, playing like a child with the little Indians, and frequently receiving with the papoose nourishment from the mother's breast. Strangely enough, the cry of the young beaver is exactly like that of the baby child. One of my friends in Michigan recently stopped at an Indian's house to see a real live baby beaver. "He cry all same as papoose," remarked the squaw, as she brought the young beaver out of the house, giving him a little slap to start him crying--and cry he did!

The body of a grown beaver is usually about thirty inches long, and something over eleven inches wide; it weighs about sixty pounds. The fore-paws are quite small in comparison with the rest of the body; the hind feet are larger, webbed like a duck's feet, and are the principal motive power in swimming. The most unique feature of the animal's body is the famous mud-plastering tail, which is oft-times a foot long, five inches in width, and an inch in thickness. The colour of the beaver varies; there are black beavers, white beavers, and brown beavers. The black are the best known.

The beaver is well equipped for defending himself, and for carrying out his architectural schemes. His jet black tail, which is like a large paddle, covered with horny scales, he uses in many ways. With it he turns the body in any desired direction while swimming and diving, and, in time of danger, employs it as a sound board, or paddle. When alarmed at night, he dives into the water, and, by means of his tail, splashes so violently as to give warning to all beavers within a half-mile distance. The stroke of the tail sounds not unlike a pistol shot. As soon as a beaver sounds the alarm all others dive underneath the water. His teeth are expressly suited by nature for cutting and chiselling out trees.

The dam is the beaver's masterpiece. In the alder or birch swamps, where he usually lives, he oft-times builds from six to eight little dams from knoll to knoll, and in this way makes a pond sufficiently large for his purposes. The average beaver dam is from twenty to thirty feet long; but they differ greatly in size. There is one on a branch of Arnold's River in Canada, where the stream is twenty-one feet wide and two feet deep, which is especially well built. The dam is seven feet high, and rises five to six feet above the pool. It is constructed mainly of alder poles, which are arranged side by side, and their length is parallel with the direction of the current. To create a pond for himself and provide against drought is the chief aim of the beaver in building his dam.

Just how these dams are built; who plans the job; who sees that it is carried out; whether each works under his own impulse or whether they co-operate; when they begin and how they finish; all these things are unknown to man. The investigation of such questions is almost impossible. It is generally believed, however, that beavers work in gangs under a common "boss" or "overseer," and it is a known fact that they work only at night. During a dark, rainy night they accomplish twice as much as on a moonlight night. No doubt the darkness gives them a sense of security which aids their work. Anyway, in the completed job, we see the evidences of a skilled engineer and architect, and one who knew thoroughly what he was about.

The size of a dam depends entirely upon the wishes of its builders and location and general conditions of land and water. Sometimes the more ambitious beavers build a dam a quarter of a mile in length. They employ exactly the same principle as is used in making a mill-dam. Beavers, however, were building dams long before millers came into existence, and their methods are fully as scientific as those of man. Mill-dams usually run straight across a stream, while beaver-dams are so curved that the water is gently turned to each side. In this way the beaver-dams are capable of resisting immense quantities of water which in its impetuous rush would carry away the ordinary mill-dam. Many scientific thinkers claim that the beaver employs this principle of construction without knowing it. How absurd! Who can be sure that he doesn't know it? Scientists of the old school desire proof before they will accept anything as a fact, yet they themselves repeatedly make wild statements without proper substantiation.

It is not unusual for a beaver family to select a home on the bank of a pond, lake, or stream whose waters are sufficiently deep and abundant for all their needs. In such a case dams are not needed, and regulation beaver houses are rarely constructed. Instead, apartment houses are hollowed out from the banks. But in the ease of a town-site on shallow, narrow waters, dams are absolutely necessary to insure sufficient depth to conceal the beavers, and to prevent obstruction by ice. The entrance to the beaver's home is almost always under the water. This arrangement safeguards the home from predatory enemies.

During the summer months, beavers are inclined to live alone, except when a new home occupies their attention; but when autumn comes, the various families of a neighbourhood meet and remain together through the following spring. In the latter part of August the busy season begins, and each and every beaver, old and young, aids in repairing the dam and dwellings, which have been allowed to fall into decay. The cutting and felling of trees is the first important work to be done.

These interesting "tree-cutters" usually work in pairs, and are sometimes assisted by younger beavers; thus the family works together in cutting and felling the trees, but in other forms of labour it seems that several families work together. If only two are engaged in felling a tree, they work by turns, and alternately keep guard; this is a well-known practice of many animals both in work and play. As soon as the tree begins to bend and crack, they cease cutting and make sure of their definite direction of escape, then they continue to gnaw until it begins to fall, whereupon they plunge into the stream, usually, where they remain for some time lest the noise of the falling tree attract the attention of enemies.

Their next work is to cut up the tree into sections which they can remove. If the tree is not too large and has already fallen in the water, they take it as it is, otherwise it must be cut up and conveyed to the dam. No professional lumberman better understands how to transport lumber to a desired place than beavers. They realise the value of water transportation and thoroughly appreciate that trees can only be removed downhill. From tame beavers we have learned that they remove smaller limbs by seizing them with their teeth, throwing the loose end over their shoulder, and then dragging them to their destination.

These water-loving animals rely mainly upon their native element for the movement of lumber and food, and to aid this they employ engineering skill that is rivalled only by their feats of tree-cutting and dam-building. This constructive faculty is shown largely in their canal-digging. From one small stream to another, or from one lake to another, they excavate canals from three to four feet in width, with a water depth of two feet, and occasionally one hundred and fifty to two hundred feet in length. The amount of labour they perform is almost unbelievable; every particle of dirt is carried away between their chin and fore-paws. This earth is sometimes used in plastering up a nearby dam or repairing their winter home. Small and tender twigs are transported to the vicinity of their lodges, and then sunk for winter food.

Mr. Morgan has made a close study of these canals, and in speaking of them he says that when he first saw them, and heard them called canals, he doubted their artificial origin; but upon examination he found that they were unquestionably beaver excavations. He considers these artificial canals, by means of which the beavers carry their wood to their lodges, the supreme act of intelligence on the part of these wise animals. Even the dam, remarkable as it is, does not show evidence of greater skill than that displayed in the making of these canals. No one who has ever understood the ways of the beaver can believe that he is not exceedingly intelligent. The banks of these canals soon become covered with growing plants and moss, and they look not unlike slow sluggish streams winding through the marshy lands.

THE BEAVER IS THE GREATEST OF ALL ANIMAL ARCHITECTS. HIS SKILL IS EQUALLED ONLY BY HIS PATIENCE.

The beaver huts, or "lodges" as they are usually called, look not unlike beehives, somewhat broader at the base, with thick walls and roof, four to six feet in thickness. They are formed of numbers of poles, twigs, and small branches of trees, woven together and plastered with mud, in the same way that the dams are made. Inside the house are circular chambers formed of mud, which have been smoothed and polished like waxed floors by the feet of the occupants. Around the outer border of each polished floor is dry grass used for Mrs. Beaver's nursery, and here the young beavers sleep and play.

From the outside these beaver huts resemble Esquimaux snow-houses, being almost circular in form, and domed. The walls are quite thick enough to keep out the cold, but with all the beaver's ingenuity, he is helpless against trappers. Summer and winter they are hunted, until now they are fast becoming extinct. How few people seem fully to realise and care what is being done to wild animals! They do not seem to know that it is a crime to take the life of a being unnecessarily. Only human life is sacred to them! To realize the wonderful work of beavers, and then to act as we do toward them is unworthy of our civilisation.

An interesting cousin of the beaver, the musquash or muskrat, and called by the Indians the beaver's "little brother," is also a house-builder and engineer of no mean abilities. He is at home throughout the greater part of North America, and, like the beaver, frequents the regions of slowly flowing streams and large, reed-bordered ponds. Here he mingles in groups of his own kin, and together they build houses, work and play, dive and swim, with almost as much skill as their big beaver brothers.

The muskrat is a skilled engineer, and delights in tunnelling. His home consists of a large rounded chamber which is reached by a long burrow from the side of a stream. From his main living-room are oftentimes found a number of smaller chambers or galleries, and these are used to store food in the form of delicate roots and bits of bark. Some of the more ambitious muskrats build large houses on piles of mud which rise out of the water. These houses are usually made of heaps of dead grass and weeds which are cemented together with mud and clay; at other times they contain no mud or clay, and seem to be only piles of tender roots and swamp grasses to be used for food during the long, cold winters.

From his physical appearance, the muskrat is well prepared to do his work: he is stoutly built, with a body about a foot in length, not including the tail; has small eyes, and tiny ears, partly covered with fur. In the winter, as food gets scarce, he begins to eat even the walls of his house, and by the time his home is gone--spring has arrived!

A most unusual family of skilled house-builders are the brush-tailed rat-kangaroos, or Jerboa kangaroos of Australia and Tasmania. They are no larger than an ordinary rabbit, but they have cousins who are as large as a man. These rat-kangaroos have most interesting tails, covered with long hair which forms itself into a crest near the tip. Their homes are found among small grassy hills, where there are a few trees and bushes. They scratch out a small hole in the ground, near a tuft of tall grass, and so bend the grass as to form a complete roof to the house, which is rather poorly constructed, and whose chief interest lies in the unusual way the kangaroos have of carrying all the building materials, like tiny bundles of hay, held compactly in their tails. There is no other workman among the animals that employs quite this method of transporting materials.

The rat-kangaroos have a dainty little brown cousin that lives in Africa, and who is occasionally seen jumping around on the ground, underneath bushes, and near damp springs. He is very small, not over three inches in length, and is like a miniature kangaroo, except for his long tail. Like their great cousins--the kangaroos--Mrs. Jerboa often carries her babies on her back when she goes out to seek food.

In the Great Sahara Desert, parched and dry, are found numerous cities of these little animals. With the exception of a few birds, reptiles, jackals and hyenas, they are the only inhabitants of this barren and desolate land. From the Arabs we learn that these little animals have extensive and intricate burrows, consisting of innumerable passages tunnelled out in the hard, dry soil. And these tunnels are the result of combined labour on the part of the entire community. The least alarm causes them to scuffle away into their underground homes.

One of the larger species of Central Asia employs a stratagem that is remarkable. Like their cousins of Africa, they live in a great underground city which is a perfect network of burrows which end in a large central chamber. From this chamber a long winding tunnel terminates very near the surface of the ground, and it is a long distance from the other burrows. No sign of its existence appears from above the surface of the earth, but if an enemy invades the burrow, away the jerboas rush for this secret exit and break through to the surface out of reach of the trouble, and escape.

These African jerboas are exceedingly odd in appearance, and they are two-legged in their habits of walk, and never go on all-fours. They walk by placing one hind foot alternately before the other; and they run in the same way. They can leap an extraordinary distance.

Frogs and toads, as a class, are not so skilled in house-building as some of their higher relations, but there is one of their number--the *Hyla faber*--that is remarkably gifted in building mud houses. He lives in Brazil, and the natives call him the *ferreiro*, or smith, and he is indeed the master-builder of his family. Mrs. Hyla is really the gifted member of the tribe, and it is during the breeding season that she diligently dives underneath the water, digs up handfuls of mud, and builds on the bottom a small circular wall, which encloses a space about ten to fourteen inches in diameter. This wall is continued until it reaches about four inches above the surface of the water. It looks not unlike a small volcano, and the inside is skilfully smoothed. This has been done by Mrs. Frog's artistic hands. When the house is entirely completed, Mrs. Frog lays a great number of eggs, and here they are quite safe from enemies both as eggs and baby tadpoles.

Mr. Frog seems little concerned in the building of the home, but he does take pleasure in croaking for Mrs. Frog while she works. Perhaps this is to her heart genuine music, and his faithful attention to their children makes up for his love of idleness!

Perhaps the strangest animal engineer in the world is found in Madagascar and Australia. It is the duckbill or duckmole, and is scientifically known as the *Ornithorhynchus paradoxus*. The natives of Australia call it by several names: *Mallangong*, *Tambreet*, and not a few call it, *Tohunbuck*.

This odd little aquatic engineer digs long tunnels of great intricacy in the bands of lazy rivers, and because of its paradoxical nature and appearance has caused many strange stories to originate about its habits and methods of propagation. It has the beak of a duck and waddles not unlike this bird, but, like other mammals, it gives birth to its young, and does not lay eggs, as is so often claimed for it. When swimming it looks like a bunch of floating weeds or grass.

Its home is always on the banks of a stream, and is always provided with two entrances: one below the surface of the water, and the other above. This insures escape in case of enemies. The main tunnel or road to the home is sometimes fifty feet in length, and no engineer could devise a more deceptive approach; it winds up and down like a huge serpent, to the right, and to the left, and is so annoyingly variable in its sinuous course that even the natives have great trouble in digging the duckbill out of its nest.

The nest is oval in form, and is well-carpeted with dry weeds and grass. Here the young reside on soft beds until they are large enough to care for themselves. There are from one to four in each nest.

There are no greater architects in the universe than may be found among the coral-polypes. These interesting little animals of the deep have been much misunderstood, and have sometimes had the erroneous designation of "insect" bestowed upon them. The word "insect" has been applied in a very loose and general sense in other days; but naturalists and scientists should see to it that the use of this term be corrected in reference to these wonderful coral-architects, and that no informed person refer to them except as animals. Even poets have been guilty of propagating the most erroneous ideas about the nature and works of these sea-builders. Montgomery, in his *Pelican Island*, makes statements that are shocking to an intelligent thinker, and which no scientist can excuse on the ground of poetical license. "The poetry of this excellent author," says Dana, "is good, but the facts nearly all errors--if literature allows of such an incongruity." Think of coral-animals as being referred to as shapeless worms that "writhe and shrink their tortuous bodies to grotesque dimensions"! These deep-sea builders manufacture or secrete from their own bodies the coral substance out of which the great reefs are built. It is a part of their life work and nature, as a flower produces its own colours and shapes; it is amusing to know that it has only been about one hundred and fifty years since it was discovered not to be a plant but an animal! Even Ovid states the popular belief of the classic period when he speaks of the coral as a seaweed "which existed in a soft state as long as it remained in the sea, but had the curious property of becoming hard on exposure to the air."

These strange coral-producing animals of the deep demand two especially important conditions only under which they will thrive: namely, a certain depth of water and a certain temperature. Thus it is seen that the warmth of the sea determines the distribution of the corals; the geography of these animals is defined by degrees of temperature. Only in equatorial seas may reef-building corals be found; and if we select the "Equator as a natural centre of the globe, and measure off a band of 1800 miles in breadth on each side of that line," we will find that it will include the chief coral regions of the earth.

The work of the corals is most interesting. Small as are these tiny workmen, each and every one does his bit and, speck by speck, adds his minute contribution to the growing mass of coral until entire islands are surrounded by extensive reefs. Tahiti, for example, is surrounded by a barrier reef which is really an immense wall. The large barrier reef on the northeast coast of Australia extends in a continuous line for 1,000 miles, and varies from 10 to 90 miles in breadth. Some reefs are mere fringes which simply skirt the coast lands, and seem to be mere extensions of the beach. Still another variety of reef is known as the "atoll" or "lagoon" reef. This latter form is seen in circular rings of coral of various breadths which enclose a body of still water--the lagoon. There are many of these coral islands in the Indian and Pacific Oceans. Keeling or Cocos Atoll, of the Indian Ocean, is 91/2 miles in its greatest width; Bow Island is 30 miles in length, and 6 miles wide; while in the Maldive Archipelago one island measures 88 geographical miles in length, and in some places is 20 miles wide. When one beholds a large coral ring, covered with rich soil and tropical vegetation, and "protecting a quiet lake-haven from the restless ocean without, it is little to be wondered at that the earlier voyagers recorded their surprise that the apparently insignificant architects of such an erection are able to withstand the force of the waves and to preserve their works among the continual attacks of the sea." As Pyrard de Laval truly said, "It is a marvel to see each of these atollons surrounded on all sides by a great bank of stone--walls such as no human hands could build on the space of earth allotted to them.... Being in the middle of an atollon, you see all around you this great stone bank, which surrounds and protects the island from the waves; but it is a formidable attempt, even for the boldest, to approach the bank and watch the waves roll in, and break with fury upon the shore."

As to the explanation of the modes of formation of these coral-reefs, the scientists have long been propounding theories which are sometimes amusing. Strangely enough they have nearly all explained that coral-polypes aggregate themselves in the forms of atolls and barrier-reefs by a mysterious "instinct," mediocrity's only term for screening its ignorance, and which is also given as the cause for their secreting lime. Flinders says that they form a great protecting reef in order that they may be protected by its shelter, and that the leeward aspect of the reef forms a nursery for their infant colonies.

Thus we see that these same scientists are accrediting these little architects with the possession of a great intelligence, and they are thought to co-operate together in a manner expressive of the greatest degree of efficiency and brotherly feeling. Each of these scientists gives a theory that leaves untouched the essential question of the causes for coral-reefs assuming their various shapes; and it is reasonable to believe that they work according to a divine wisdom and plan, and that mankind does not yet understand their strange ways, which give us a higher conception of the universe than that held by the ancients. Science has come to the point where it must recognise the perfect unity of all life, and that our fellow-architects, engineers, and house-builders in the animal world also fill an important place in Nature's great scheme

XI

FOOD CONSERVERS

"He prayeth well who loveth well
Both man and bird and beast.
He prayeth best who loveth best
All things both great and small;
For the dear God who loveth us,
He made and loveth all."

--*COLERIDGE.*

It can almost be said that there is no industry or profession of the human world that is not carried on with equal skill in the animal world. This is especially true of merchandising and store-keeping; animals, however, have different methods of merchandising than men, although these methods are none the less real. They give and take instead of buy and sell and have co-operative shops which they operate with great success. They unite for a desired end, and demonstrate their ability to work together in a common enterprise in a way that might teach man a good lesson.

Food and shelter are the first needs of animals. In order to obtain these, they group themselves into foraging parties in the most ingenious manner. Like mankind, they sometimes co-operate for dishonest ends; they form "trusts" and organise into gangs for purposes of mutual aid.

Deer, monkeys, rabbits, foxes, and numerous others conduct their dining-rooms on a co-operative principle. Some watch and wait while others dine. The same is true where they go to watering places to drink and bathe.

Perhaps the most unique and clever food conserver is the American polecat. He not only provides for himself, but prepares a larder for his young, so that they will have plenty of food. The nursery is usually comfortably embedded in a cave, and is lined with soft, dry grass. Adjoining this nursery is a larder, which often contains from ten to fifty large frogs and toads, all alive, but so dexterously bitten through the brain as to make them incapable of escaping. Mr. and Mrs. Polecat can then visit or hunt as they please, so long as their children have plenty of fresh meat at home!

Another interesting food conserver is the chipping squirrel, or chipmunk, so named because his cry sounds like the chirp of little chickens. His method of dress is most unusual; he is brownish grey in colour, with five stripes of black and two of pale yellow running along the back of his coat; the throat and lower part of his body is snowy white. These colours occasionally vary, when the grey and yellow are superseded by black.

His home is underground, usually under an old wall, near a rock fence, or under a tree; his burrow is so long and winding that he can easily escape almost any enemy, except the weasel, which is not easily outwitted. His nursery and living-room is quite pretentious, but his lateral storeroom is a marvel! He is a miser indeed, and stores up every acorn and nut he can find, even many times more than he can ever eat. His variety of food is almost unending--he loves buckwheat, beaked nuts, pecans, various kinds of grass seeds, and Indian corn. In carrying food to his home he first fills his pouches to overflowing and then takes another nut in his mouth; he thus reminds the classical reader of Alemaeon in the treasury of Croesus.

The hedgehog is a regular Solomon in her methods of collecting fruit. Plutarch had a very high opinion of her. He says that when grapes are ripe, the mother hedgehog goes under the vines and shakes them until some of the grapes fall; she then literally rolls over them until many are attached to her spines, and marches back to her babies in the cave. "One day," says Plutarch, "when we were all together, we had the chance of seeing this with our own eyes--it looked as if a bunch of grapes was shuffling along the ground, so thickly covered was the animal with its booty."

American Museum of Natural History, New York

THE SKUNK MOTHER TRIES TO KEEP ON HAND A GOOD SUPPLY OF SUCH DELICACIES AS FROGS AND TOADS, SO THAT HER YOUNG MAY NEVER GO HUNGRY.

American Museum of Natural History, New York

THE PORCUPINE AND THE HEDGEHOG HAVE A UNIQUE METHOD OF COLLECTING FOOD FOR THEIR YOUNG. AFTER SHAKING DOWN BERRIES OR GRAPES, THEY ROLL IN THEM, THEN HURRY HOME WITH THE FOOD ATTACHED TO THEIR QUILLS.

Alpine mice not only form comfortable winter homes in the earth, but combine into small winter colonies, each colony numbering about ten to twelve inhabitants, all of whom are under the direction of a leader. Thus organised, they proceed to lay up provisions for the winter. They use their mouths as scythes and their paws as rotary machines. Surely their wisdom and foresight call forth our greatest admiration. The jerboas or jumping mice are not only skilled athletes in the art of jumping, but they are gifted food conservers and producers as well. They lay up complete storehouses of food, which they do not consume altogether as their appetite may direct; but conserve it carefully for the times when nothing can be obtained from the fields. Then, and then only, do they open the closed magazines. Such acts of intelligence cannot be recorded under the head of "instinct"! They demonstrate the ability to plan for the future, and meet all emergencies.

Certain food hoarders and robbers, like the vole, are so very greedy and become such misers that they often threaten total destruction to large areas of grain. They were so plentiful in the classic land of Thessaly, the vale of Tempe, and the Land of Olympus that the old Greeks established what they called an Apollo Smintheus, the Mouse-destroying God. In the early spring, according to Professor Loeffler, who has made a special study of their invasions, they begin to come down from their homes in the hills to the cultivated fields. They seem to follow regular roads, and often travel along the railroad embankment. They travel very slowly, and when at home live somewhat on the order of prairie dogs, that is, in underground dwellings with numerous winding passages and tunnels.

These wise little food conservers are nocturnal in habit, and are rarely seen except by careful observers. When they once determine to rob a field, they do it with amazing rapidity and completeness. In a single night hordes of these workers go into a cornfield and by daylight not a stalk of corn remains. The field is as empty as if a cyclone had struck it. They work with great system, and while a part of their number cut the stalks down, others cut it up into movable sizes, while still others superintend its systematic removal. Storehouses are usually provided before the grain is even cut. They make long voyages throughout a country, storing away tons of grain and food in these various granaries. To these they come for supplies whenever necessary. All poverty-stricken voles are also fed from these storehouses, since it is the product of the community as a whole. Aristotle wrote at length about their wise and destructive ways.

Not the least ingenious of food conservers are the hamsters, members of the great rodent family. They have made their dwellings most comfortable and even luxurious in arrangement and furnishings. Like wealthy farmers, they are not satisfied with comfortable dwellings only, but they too must have spacious barns adjoining their homes. Their home, or burrow proper, consists of two openings: one, which is used as an entrance, and which sinks vertically into the ground; the other, which is used as an exit, with a winding slope. The central room is beautifully carpeted with straw, moss, and dry leaves, which makes it a very pleasant living-room and bedroom. A third small winding tunnel leads from this room to the barns and storehouse. Thus, Mr. and Mrs. Hamster and the children have no need to go forth in the cold and wet weather to seek food--they can remain at home perfectly protected and well-fed. They are very liberal, and in case of need or poverty, will always share their food with their neighbours.

I once found the nest of a harvest mouse, which was woven of plaited blades of straw of the oats and wheat. It was perfectly round, with the aperture so ingeniously closed that I could scarcely tell to what part of the nest it belonged. It was as round as a marble and would actually roll when placed on a table, although within its walls were six tiny mice, naked and blind. As they increased in size day by day, the elastic wall of their small home expanded, and thus served their need until such time as they were old enough to live independent of this specially provided shelter.

There is a larger animal, known as a "rat-hare" or the harvest rat, which gathers piles of hay for winter use, sometimes to the height of six or eight feet in diameter. They begin harvesting in the early part of August, and after having cut the grass, they carefully spread it out to dry before placing it in their barns. These barns are usually located in holes or crevices of mountains. They are found in immense numbers in the Altai Mountains.

The California woodrat is not only a food hoarder but a notable thief and robber. A nest was found that was a veritable tool chest and pawn shop! It contained fourteen knives, three forks, six small spoons, one large soup spoon, twenty-seven large nails, hundreds of small tacks, two butcher knives, three pairs of eye-glasses, one purse, one string of beads, one rubber ball, two small cakes of soap, one string of red peppers, several boxes of matches, with numerous small buttons, needles, and pins. Apparently these woodrats are as ambitious for unnecessary and useless possessions as is man himself. Their big storeroom did, however, contain a larder in which they had some of their favourite food, such as seeds and nuts.

Some animals have learned not only to acquire, but also to defend and protect, all their property. We see in the human world how strong is the impulse to collect, and children will invariably collect anything from pebbles to peach-pits, if they see other children doing the same thing.

Most animals that do not hoard are those that forage for food, or fish, and rarely have permanent homes. The orang-outangs, for example, are regular gipsies, and go from place to place wherever food is plentiful. They take life easy, and sometimes during their journeys select a suitable spot near the seashore and have a real picnic. A scout has already discovered the right spot for getting big oysters, of which they are exceedingly fond, and when they have assembled, certain ones proceed to dig up the oysters, which they hand to others on the shore and they, in turn, place them on big stones, and proceed to open them for the feast. If one of the fishermen-monkeys discovers an oyster open, he will not insert his hand to remove the meat until first placing a stone between the valves. This assures him protection against the closing of the oyster. In most cases, they open the oysters by first placing them on stones and then using another stone as a hammer. These facts are vouched for by no less authorities than Gamelli Carreri, Dampier, and Wafer.

It is only a matter of time until many animals will understand the use of man-made tools. Some have already learned to use such tools as they make and shape for themselves. Monkeys and apes are already gifted in this art. Of course, under domestication, they use knives, forks, spoons, and dishes not so much from intelligence as from imitation. This, however, might be said of many human beings. I have seen an immense chimpanzee sit in a chair, set his own dinner table, use his knife and fork correctly when eating, and take great delight in the use of his napkin, which he always carefully refolded when his meal was over.

The human-like qualities of apes and monkeys, however, need scarcely be told. They are so very similar to man in most ways that there are few things they cannot do. Aelian tells of an ape which learned to drive horses skilfully. He knew just when and how to use the whip, how much slack to allow in the reins, and when to tighten them! They greatly resent any intrusion on their hunting-grounds, and make use of sticks and clubs to protect them. The chief is always armed with a club, and is thoroughly skilled in the use of it. It sometimes happens that an elephant will come to the same tree to seek food that apes frequent, and although they have no enmity towards each other, they like the same kind of food. As soon as the ape sees the elephant reaching his trunk among the branches, he immediately slips near the elephant, and when an opportunity presents itself, he whacks him over the trunk with his club! The infuriated elephant runs away in terror!

A story is told of a party of foraging apes who went into a cornfield with the purpose of robbing it, and discovered two men. They immediately rushed upon them and attempted to poke their eyes out with sticks and would have succeeded but for the intervention of two other men who chanced to be near. The extreme cleverness of apes in applying their reason and judgment is shown in Vosmaer's account of the female orang-outang, who tried to open the padlock of her chain with a small stick. She had seen her master open it with a key, and she exactly imitated the motion of his hands in the attempt.

Man shows a disposition to deny animals all traits and characteristics which are similar to his own. This reminds us of a remark that Cardinal Newman once made that men know less of animals than they do of angels. Why should we show such foolish pride and delusion, and try to baffle one of God's great facts? When men attempt to extinguish the idea of animal intelligence and sentiment by referring to it as instinct, we are reminded of the desert ostrich, which buries its head in the sand and thinks it cannot be seen. We should proudly acknowledge the wonderful human-like methods of these food conservers of the animal world, and recognise in all this a guiding Providence who provides for and protects all his creatures, be they great or small.

XII

TOURISTS AND SIGHT-SEERS

"Every night we must look, lest the down slope
Between us and the woods turn suddenly
To a grey onrush full of small green candles,
The charging pack with eyes flaming for flesh.
And well for us then if there's no more mist
Than the white panting of the wolfish hunger."

The desire to travel and see the great world is by no means peculiar to the human race. It is found among animals to such a degree that groups of them will often leave their homes in one country and journey to another. These strange wanderlust habits are noticed even by the casual observer, and no special insight is required to see that these wise creatures have their annual tours excellently arranged and marked out. Their route is possibly as definitely arranged before starting, as is the route of a human traveller. They have their selected eating places arranged, know every danger spot and the enemies they are likely to encounter.

The members of these co-operative tours take life tickets, and each tour lasts about one year. One of the most unusual instances of such co-operation is that of the lemmings of the Scandinavian countries. These are animals of the mouse tribe, which live in the mountainous districts. They live upon roots and grasses. They breed very rapidly. At certain times they go from the centre of Norway to the east and west, crossing valley, hill, and river in great masses. Many are destroyed by birds and beasts of prey, but finally the survivors reach the Atlantic on the Gulf of Bothnia and, for some strange unknown reason, plunge in and die. Only enough remain from one season to another to propagate the species. It is an immense co-operative suicide society.

Rivers and valleys are sometimes effectual barriers. On the plains of the Amazon great numbers of animals are found on one side of the river only; these have not been able to cross to the other. On the north side of the Rio Negro are two varieties of monkeys, the *brachiurus conxion* and the *jacchus bicolor*, which are unknown on the south side. Of course, water-loving animals, such as seals, whales, and porpoises are at home in the water and can swim for days without stopping. Quite a few animals can swim for a short distance, but comparatively few for long distances. In the early days in North America it was not uncommon for buffalo to swim across the Mississippi River. Rats and squirrels often migrate in great numbers. It oftentimes happens that Arctic animals travel from one place to another on floating ice. In the South American waters it is a common sight to see floating islands covered with plants and trees upon which there are live animals; and while these animals are likely to perish, they are oftentimes carried safely to land. Eagles have often been instrumental in bringing new species of animals to islands where they had previously been unknown, their purpose being to provide food for their own young. Some of these animals would escape and henceforth become citizens of their new habitation.

An interesting division of migrants is that of the casual travellers, like the men and women who always remain at home except when special business calls them away. Sudden climatic changes, or the scarcity of food, often cause stay-at-home animals to make tours into new territories. As a good instance, I might cite the case of three wolves, which I saw entering Jackson Park in Chicago, during very severe weather when Lake Michigan was frozen over. The morning papers stated that because of forest fires in Michigan, and the extreme cold, which not only made food scarce for the wild animals of Michigan, but froze the Lake, many of them had come across the ice into the great Chicago parks seeking food and shelter.

The subject of animal travel is full of interesting and difficult problems, and not the least interesting nor the least difficult is the question of just how they find their way to and from various places. Many naturalists tell us that these animals are led by inherited instinct along the migration lines followed by their forefathers. But even if this were true, what made them originally follow such a course?

Wild horses when travelling always have a leader as well as several sentinels for each herd. By some unknown code this leader makes known his wishes and directs the movements of the herd. No human army could have greater order or more perfect obedience to commands; and under him there is absolute unity by means of which the carnivorous animals, such as the wolf, the jaguar, and the puma, are repelled. Wild deer invariably have a leader, and while we do not know how he obtains his position, nor how he directs his followers, we do know he is highly successful in his efforts.

No act in the animal world bespeaks more intelligence than that of placing sentinels, especially during a journey. Horses show striking skill and ingenuity in the choosing and placing of their sentinels. Any one who has been fortunate enough to have seen them travelling in the forests of South America, where the wild horses are gregarious, and travel in herds of five hundred to a thousand, has noticed that sentinels are always stationed around the herd. These animals are not well prepared for fighting, and experience has taught them that their greatest safety is in flight, and so, when they graze or sleep, sentinels are always on the look-out for enemies. If a man approaches, the sentinel at first walks toward him, as if to make sure what the enemy is, and what he desires, if the man goes nearer to the herd, the sentinel neighs in a most peculiar tone. Immediately the herd is aroused, and gallops away, not in confusion, but perfect order, as though its members were human soldiers.

The same is true of the white-legged peccaries, so plentiful in Guiana. They congregate by the thousands, choose a leader whose position is always at the front, and travel for hundreds of miles through the great forests. If they come to a river, the leader halts, as if to make sure that all is well for crossing, then he plunges into the water and is followed by his immense army. The sureness of the leader would suggest that he has been over the same route many times before--perhaps this is why he has been chosen! If an enemy appears, or any form of danger is approached, they carry on an immense amount of chattering and proceed only when they have talked it out. Any hunter that should be foolish enough to attack them, unless he were already up a tree, would be torn to pieces with their terrible teeth and tusks. They are as bloodthirsty as the wild boars of the Black Forest of Germany, and will sometimes actually tear down a tree up which an enemy has escaped, that they may kill him.

The African apes have an interesting way of sending their sentinel to the top of an adjacent rock or tree, that he may look over the surrounding valleys and plantations before they go to plunder a garden or field. If he sees any danger, he utters a loud shriek, and the entire troop immediately runs away. The monkeys of Brazil post a guard while they sleep; the same is true of the chamois and other species of wild antelope.

A few years ago, many of the sheep in the northern part of Wales had become quite wild, and they usually grazed in parties of twelve to twenty, always having a sentinel so stationed as to command a prominent view of the surrounding territory. If any animal or person came near, he would give a peculiar hiss or whistle, repeating it two or three times, at which the whole herd would scamper away to places of safety.

One of the most striking facts about migration is its never-failing regularity and success. Most animals migrate at the recurrence of the breeding season. Of these, the great sea-turtle, which seeks the shallow water and deep sandy hills when ready to lay her eggs, is well known. Notwithstanding the great risks that practically all travelling animals assume, they are successful as a whole in their travels, and many return to bear testimony to a successful trip even across continents and sometimes the ocean. They migrate, for a variety of reasons. When it is not for a more desirable climate, nor more food, nor even better breeding grounds, we must either believe it is because of the natural desire to travel, or frankly admit that we do not understand it.

The Icelandic mice have probably the most curious methods of travelling of all migratory animals. Dr. Henderson, an authority on Iceland, not only verifies the fact himself, but gives the names of many prominent investigators who have seen the mice crossing small rivers and streams on thin pieces of dry board, dragging them to the water, launching them, and then going aboard their little rafts. They then turn their heads to the centre, and their tails, which hang in the water, are used as paddles and rudders until they reach the destined shore.

Among travellers none are more famed than the camels. In their sphere and use they are supreme, and Nature has prepared them especially for travelling on the dry, hot, and barren deserts. They are truly the "ships of the desert" for they travel on a sea of sand, and their pad-like feet, so poorly adapted for travel on moist soil, is admirably suited to the desert sands. They are capable of travelling many days without food or water, and are used extensively in the desert regions of the East not only as beasts of burden but for their milk, which is an important article of diet in those countries where the camel is at home.

Animals that do not migrate, especially those living in cold climates, change their clothing at regular intervals. Their hair or fur increases in thickness in winter. If we compare the Indian and African elephants of to-day, whose delicate thin hair is scarcely noticeable, with the great extinct mammoth, which had an enormous amount of woolly fur, we readily see the great difference in their clothing. Yet these animals are members of the same great family. The same difference may be noted with horses: the Arabian horse, for example, has short, glistening fur, while those of Iceland and Norway have very thick fur; the same is true of Northern and Southern sheep. Animals which live in temperate regions, put on much thicker coats in winter, and shed them as summer approaches.

American Museum of Natural History, New York

THE BLACK BEAR IS NOT ONE OF THE GREAT MIGRATING ANIMALS. THE THICKNESS OF HIS COAT MUST THEREFORE CHANGE WITH THE SEASONS.

American Museum of Natural History, New York

RABBITS SEEM TO HAVE A WELL-DEVISED SYSTEM IN THEIR ROAD-BUILDING, RUNNING THEIR PATHS IN AND OUT OF UNDERBRUSH IN A TRULY INGENIOUS MANNER.

The love of their original homes is one of the most striking features of certain animal travellers. The fierce struggle for existence and the territory required for an animal's home largely determine the amount of effort they make to seize and hold certain possessions. A pair of wildcats, for example, require a comparatively small hunting ground. But this they will defend against invasion even to the point of death. There are many more evidences showing the animals' love of home, and that they also know the meaning of home-sickness.

Not a few animals have learned definitely to lay out and obtain recognition for the boundaries of their respective ranging-grounds. This is amply proven by their respect and recognition of rights of way. Animals of certain farms seem to know the exact boundaries of their grazing lands and pastures, and to teach this knowledge to their young.

In addition they often police their lands and pastures against intruders. Woe unto any traveller found on the wrong highway! It is not uncommon for the transgressor to be pushed from a right of way to the rocks below. More than once a court's decision regarding disputable territory has been based on the sheep's recognition of boundary; those sheep slain in battle or otherwise injured while trying to invade the questionable territory have been paid for by the owner of the transgressing sheep.

It is easy to understand how sheep can recognise their rights of way, but somewhat difficult to account for their knowledge of boundaries. Sheep and goats have for ages been the greatest mountain-path and road-makers. Whether or not they have engineers, we are not sure, but they seem to select the shortest, easiest, and best route across the trackless hills, and never seem to change the way. In these localities, the sheep are almost in a primitive condition, and "not the least interesting feature of their conduct in this relapse to the wild life is that, in spite of the highly artificial condition in which they live to-day, they retain the primitive instincts of their race."

That this "peremptory and path-keeping" instinct is shown by the habits of the musk-ox, is clear. He is as much akin to the sheep as to cattle, and in habits more like those of the great prehistoric sheep as we imagine these to have been. The musk-ox naturally assembles in large flocks, and is migratory, just as the domesticated flocks of Spain are, and those of Thrace and the Caspian steppe. These flocks always return from the barren lands in the far north by the same road, and cross rivers by the same fords. Nothing but too persistent slaughter at these points by the enemies who beset them, induces them to desert their ancient highways. Pictures and anecdotes of the migrations of these animals, and of the bison in former days, represent them as moving on a broad front across the prairie or tundra. The examples of all moving multitudes suggest that this was not their usual formation on the march, and their roads prove that they moved on a narrow front or in file. On the North American prairie, though the bison are extinct, their great roads still remain as evidence of their former habits. These trails are paths worn on the prairie, nearly all running due north and south (the line of the old migration of the herds), like gigantic rabbit tracks. They are hard, the grass on them is green and short, and, if followed, they generally lead near water, to which a diverging track runs from the highway.

How interesting must have been the life on this great animal highway, before the Indian made the deadly arrow to destroy these nature-loving travellers! There is no doubt but that, in their own way, these animals felt all the emotions known to a human traveller; that they enjoyed the flowery road, rested and played when weary, looked forward with joy to their favourite watering and bathing places, and recognised old watering places that they had visited for years.

The great roads and highways made by graminivorous animals, from those which the hippopotamus cuts through the mammoth canes and reeds of the African streams, to the smaller rabbit highways of England and America, all tell their own story of how these animals live and travel. The principal roads of rabbits over hills are as permanent as sheep and buffalo roads. These roads, however, should not be confused with the little trails that lead to their play and feeding grounds.

My friend and fellow-naturalist, Ralph Stuart Murray, in writing to me from Quebec, says: "In speaking of animal road builders, I might say that the rabbit or hare of the north woods deserves much attention, for greatly interesting are his highways. The life of the north woods brings one constantly in touch with these roads, which, after generations upon generations of constant use, are worn deep and smooth into the moose grass and muskeg through which they run. At places, several distinct paths intersect, and it is curious to note that while these roads wind in and out underneath the low hanging evergreens, the 'cross-roads' will invariably be located in a clear open space, often on the top of some small hillock.

"The great age of these roads is very evident when compared with the newer, shallower paths of more recent years. So deep are the old ones, in fact, that the quiet watcher in the woods will occasionally see two large, upright ears--unmistakably those of a rabbit, seemingly sticking out of a hole in the ground--yet moving at a rapid pace, and all the while no rabbit in view. For all the world these vertical ears belonging to an unseen owner resemble in use and appearance the periscope of a submarine--the difference being that the rabbit uses his 'periscopes' for hearing, in order to locate and avoid his foe, the submarine its periscope to locate and attack its enemy."

The sheep terraces, which are so common on the sides of hills, though made by sheep, are not roads, but feeding grounds. Sheep, when walking on a hillside, invariably graze on the upper side, as they cannot reach the lower grass. Therefore they walk backwards and forwards on the slope, just as a reaping machine is driven over a hillside wheat-field. As the sheep takes a "neck's length" each time, the little ridges or roads correspond exactly with the measurements of the sheep's neck.

There are as many kinds of roads and terminals in the animal world as there are in the human, and lest our pride make us forget, we should remember that even the Panama Canal is dug according to the plan of a crawfish's canal, such as may be seen near any muddy stream. It is strange that no animal has learned to build elevated roads, though animals that live in trees, like flying squirrels, monkeys, and flying foxes, are very skilled in going from one tree to another. They have regular aerial highways, and some of the tree frogs are veritable wonders in the accuracy of their leaps from tree to tree. Even more skilled than these are the agamid lizards of India, whose chief means of travel is a folding parachute, which at a moment's notice can be erected and carry to another tree its lucky possessor. In Borneo is an aviator tree-snake which is able to so spread his ribs and inflate his body that he can actually sail from branch to branch in the tree-tops.

There are night travellers as well as day travellers; in fact, there are more animals that roam around in a great forest at night than in the daytime. They sleep during the day, when the day animals are roaming about, and go forth to roam when it is night. It is then they seek for prey, and are much feared by day animals. They see well in the dark, and travel so lightly that their footsteps cannot be heard.

On the Island of Java are found a family of strange, dwarfish little beings, which are called by the natives malmags, or hobgoblins. And they are well named, for they look like creatures of a distorted imagination more than real, living animals. They travel only at night, and so superstitious are the natives of their evil influence that if one of these uncanny little creatures appears near their rice fields, the plantation is immediately abandoned. However, these small creatures are no larger than squirrels, and are perfectly harmless. They are very rare even in their native lands--the Oriental Archipelago and the Philippine Islands. They rear their young in the hollow roots of bamboo trees, and to disturb their nests means to incur the evil of all the land.

Night animals do not go forth to travel and seek prey until the night is far advanced, and their prey is soundly sleeping. They seem to know the exact time of the night, as if they had watches or clocks, and they usually go forth to hunt about midnight and return to their homes about four o'clock. Only in cases of extreme hunger do they vary from this rule.

How marvellously skilled are they in finding their way! They pass through a crowded forest as though it were daytime, and strangely enough know just how to return to their lairs. This special sense or gift is not possessed by man; he must have marks and signs to return to a definite place.

These night-travellers number among their lot bats, flying squirrels, leopards, and prowling snakes.

Bats are not only the most interesting of the night-travellers, but by far the most curious and wonderful animals in the world. They are hideously ugly, reminding one more of a miniature, closed-up umbrella than an animal! They are coarse, awkward, when not in flight, and repellent; yet they have such highly developed senses that they have no rivals in the animal world. They excel most birds in flight, are able to make long nightly journeys, in which they use their wings not only for flight, but as air-bags in which they catch all kinds of flying insects. Their sense of touch as we know it is really a combination of touch, sight, and hearing.

A bat is a paradox par excellence! Nature seems to have started to make a little bear or fox, and suddenly forgot how and changed it into a winged freak, with tail, claws, fur, sharp teeth, small ears that stand up, and tiny, half-buried eyes. Its queer angular-edged wings look like an umbrella, with the cloth stretched over steel ribs; but in the case of the bat, this framework is made of delicate bones which are covered with a thin skin. The skin contains numerous little sense organs dotted over its surface, which give the bat his strange power.

Bats look more like mice than they do like birds, and they are sometimes called flittermice. But they are mammals, and the young are fed with milk by the mother, just as a cow feeds her calf. There is no danger that a bat will ever fly against you in the dark; for they can avoid all mishap even when their eyes are put out. They have special sense organs that tell them when they are nearing an object, and can fly at headlong speed with the accuracy of a rifle bullet directly into a small opening. This power is all due to the mysterious sense located in their wings and ears, which causes even man to consider his senses weak in comparison.

Bats are sociable creatures and huddle together and sleep in vast numbers during the day, but when night comes on they come forth for their nocturnal travels and sport by the millions. I have seen them leaving caves just at dusk in such numbers as to look like one immense volume of smoke, twenty to thirty feet wide, and lasting for more than five minutes. Mrs. Bat often takes her babies with her on these nightly travels. I found one with two young clinging to her breast. How they must enjoy these lovely trips!

There are many kinds and varieties of bats, ranging in size from the flying foxes of the tropical world, with wings five feet in length, to the wood bat of North America, which is not over six inches long. These interesting friends of man are his greatest scavengers of the air. They are doing much to check the mosquitoes throughout the regions of the world, and in more civilized communities man makes shelters for them, that they may eradicate mosquitoes.

XIII

ANIMAL SCAVENGERS AND CRIMINALS

"A warning from these pages take,
And know this truth sublime--
Each creature is a criminal
When he commits a crime."

No more remarkable creatures exist in the animal world than those that play the role of Nature's scavengers and criminals. They are as numerous and varied in their methods of working as they are interesting. The only things they have in common are their profession and their appetites. As individuals they are ugly, unattractive and apparently void of personality and charm. Nevertheless, they have an important part to play in the scheme of things.

One of the most noted of these scavengers is the jackal--the Bohemian of the desert--whose territory extends from the Gulf of Persia to the Strait of Gibraltar. He is equally at home in Arabia, Persia, Babylonia, Syria, Egypt, and the entire North Coast of Africa, and no country from Barbary to the Cape of Good Hope is ever out of reach of his ghostly and uncouth howls. He travels only by night, and very rapidly.

When suffering with extreme hunger, he will attack man, but this he will do only in very rare cases. As he lives entirely upon dead animals, he is more of a thief and glutton than a robber and murderer. He depends mostly upon flight and darkness for his protection, and rarely ventures a direct attack. With all his unlikable habits he is truly valuable as an agent of public salubrity, and an important officer of the desert "commission of highways."

These public scavengers, while especially fond of carcasses and putrid flesh, are not averse to a little fresh meat occasionally. The jackal is truly the follower or purveyor for the lion, and oftentimes they work together. Jackals will gather in large numbers near a lion's den and howl and scream until the lions come forth to disperse them. As soon as a lion appears they stop their noise, but when he is out of sight, they immediately begin again. This is done because game is near, and the wise jackals wish the lion to kill the game. When this is done, and the lions have eaten all except the bones, the jackals have their small feast of scraps.

These weird night prowlers have ways all their own, as any one who has spent a night in a tropical desert can attest. Imagine yourself on the Syrian plains between Bagdad and Damascus; a small white tent, and a starry sky: the silence is appalling, and you are just about to have your first sleep in the desert. Away, away from the distance comes a mournful, ghostly cry. Suddenly it ceases and like myriads of echoes it is repeated in hideous intensity--a babel of cries weird beyond description--so fierce and screeching as to be almost blood-curdling. It seems to come from all directions and distance out of measure! Vibrating over the sands and through the rocks, filling the immense void, crying out as it were for the sphinx, a veritable *de profundis* of the wastes. The vultures, who hold the fort during the day have given way to the night shift, the jackals. These come from all directions; from the caves in the earth, from among the rocks, from here, there, and from everywhere to take up their hygienic services where it has been left off by the day scavengers.

If you were near an oasis in the desert at the close of day, you would suddenly hear from the hot, barren sands a deep and peculiar sound. It swells and grows as an approaching wind, growing louder and louder as it comes nearer. Suddenly by the light of the camp fire, you see myriads of horrid green eyes, like ghost torches in a graveyard, and hear gnashing teeth, greedy in anticipation of the garbage you have thrown away.

These hyena hordes are frightfully ugly, but rarely dangerous to man. They visit every oasis settlement in immense numbers, howling, yelping, and fighting for any bit of offal they may find. Not a particle of garbage remains. At the first sign of dawn, they disappear like rats from a burning building, and seek their caves to digest their ignoble banquets.

No human street-cleaner could ever excel their work. No matter how large the garbage pile, no matter how many dead dogs, cats, and donkeys in a village street, no matter how unspeakable the offal, it all vanishes as completely as though it had been burned. Not a piece of bone, not a single chicken feather remains. The natives have no fear of the hyena; a small child armed with a stick can put to flight a dozen of them. They are the lowest of cowards, and will flee from their own shadows.

THE MONGOOSE IS A SCAVENGER OF THE WORST TYPE, FEEDING ON RATS AND MICE AND SNAKES, AND EVEN POULTRY.

American Museum of Natural History, New York

DIPLODOCUS. THE PREHISTORIC ANIMALS, ALSO, UNDOUBTEDLY HAD THEIR SCAVENGERS AND CRIMINALS.

In spite of their valuable services, mankind hates the hyenas. This is probably because of their absolute cowardice, for they will never attack a living creature unless it is weak from illness. Sometimes they steal a baby, never killing it outright, but carrying it away to their dens to starve it to death before mutilating its body. If the courage of this beast equalled his strength, he would be the despot of the desert. But he is like his fellow workman, the jackal, cowardly to the last degree.

Neither of them ever attempts to put an enemy to flight by legitimate means. They resort to fakery: one howls, and the other wrinkles his face in great anger. The jackal's greatest asset and protection, when he meets with an enemy, is bluff. He raises his ugly mane, lifts his ungainly shoulders and assumes the look of a Jason, while in reality he is as harmless as a mouse, and the smallest child could drive him away with a twig. His bravery is all pose--a make-believe game--which he plays over and over again with every one he meets.

A noted American scavenger is the peccary, a species of wild hog, whose home ranges from Texas to the Pampas of South America. He is a devourer of creatures more obnoxious than himself. He moves with great rapidity, is always on the alert, and stops at nothing from mountains to a flowing river. When he attacks an enemy he makes short work of him.

Bands of these hogs are led by a chief, who is the swiftest and fiercest of the herd. This aggressive leader is followed by successive lines of males, behind which come the strong females, while the rear is brought up by the old, the sick, and the young. In marching, they have the discipline of a trained army, and turn neither to the right nor to the left but go straight ahead. If the leader, for any cause, decides to change his route, the fact is quickly made known in some way to his followers, and the turn is made at a direct angle, with the accuracy of a surveyor, and the peccaries go forward again directly toward their new destination. This is another evidence of a special sense unknown to man.

But whenever a stop is made, or wherever they go, they do their work as scavengers. Fallen fruits, dead animals, insects, snakes, and worms are their prey. Thus they are valuable forest sweepers.

Strangely enough, in the animal world, as in the human, the lower professions are filled with those of less mentality than the higher, and as a result we find scavengers are nearest allied to criminals. The idea of one creature killing and eating another seems terrible. Yet they do, and most often do human beings commit the same crime. Cannibalism among wild animals is a common occurrence. The demand for food usually causes one animal to kill and devour another. But in captivity there are other causes for cannibalism: fear and excitement will oftentimes cause a mother to destroy her offspring.

It is a case of dog eat dog! Badgers often kill and devour their young. Wolves, in cases of extreme hunger, will eat their puppies; and Arctic travellers, when food for their dogs is scarce, have to guard constantly against the stronger eating the weaker. I once caught a mother field mouse with her two young and placed them in a cage; the next day the young had strangely disappeared, but I am not sure that the mother had eaten them. Hogs, cats, and rabbits will sometimes kill and eat their young even when food is plentiful. Crocodiles show an occasional cannibalistic tendency, while water-shrews are very pugnacious and oftentimes fight until one is killed. The victorious one eats his enemy!

Thus it appears that Nature does not entirely disapprove of cannibalism, or she would not allow so many of her creatures to practise it.

Theft is a common vice among these various criminals. Monkeys and baboons form regular bands to rob and plunder. They have a chief who sees that a sentinel is posted at each dangerous post. The plunderers then line up in a long row, and the leader gets the booty and passes it along the line until it reaches the last of the band--the receiver. He deposits it in a safe place. If the sentry sounds an alarm, they all flee away, each with as much booty as he can grab. If the enemy presses too close, all booty is thrown away.

Passion, especially of love, causes much crime among animals as it does among men. Jealousy burns fiercely even in the breast of a beast. It is a common heritage of the fiercest lion and the gentle gazelle alike, and is capable of perpetrating the most dreadful crimes.

There are types of ugly dispositioned animals, who are always in a ferocious mood, just like certain ill-tempered human beings, who believe everything and everybody is trying to injure them. The common shrew, for example, is noisy, bold and fussy. He seems to delight in calling attention to himself by his grunty, squeaky voice. He advertises himself as a bad animal; and bad he is, for his terrible odour prevents other animals from coming near. Horses and mules are at times quite ferocious, and kick and bite, with no idea of obedience or kindness. They, of course, like our human criminals, are mentally unbalanced. Skilled horse trainers can detect at a glance a criminally inclined horse.

Rogue elephants are common in India. Even their trumpeting shows a ferocity and unbalance that terrifies the natives. Often these criminal elephants are sufferers of mental ailments. A respectable, law-abiding elephant herd will not allow a thug or rogue to live in their midst. They recognise him as dangerous for their society, and combine to force him entirely away from their homes.

Certain criminal animals have a strange antipathy for members of their own tribe, or for other kinds of animals. Such is common among monkeys, cats, horses, and dogs, and many terrible crimes are committed because of these antipathies. Every one has witnessed the terror of a dog that has been insulted, and elephants will carry an old grudge for fifty years and finally seek the most terrible revenge.

Often violent outbursts of temper on the part of a tame animal are caused by a change in the temperature or atmosphere. Even animals have days when they feel ugly and grouchy. Those that live in very hot climates are especially subject to fits of rage and anger. The approach of an electrical storm causes many of them to lose their self-control: herds of cattle often stampede just preceding a cyclone. They, like human savages, seem terrorised at the unknown. Not a few wild animals have actually run in the way of an automobile or passing train to attempt to stop it. Fear and rage are often caused by the appearance of a curious object. A bull, for example, when he sees a red rag, will madly rush at it, seemingly altogether oblivious of the man holding it. The matadors are safe only because the bull is insane from rage.

Many scientists of fame, like Lombroso, have demonstrated that strong drink is the cause of much crime among animals, the same as it is among men. In the pastures of Abyssinia the sheep and goats get on regular "drunks" by eating the beans of the coffee plants. They fight and carouse at such times like regular topers. Elephants are incorrigible when drunk, while dogs and horses have to be put in strait-jackets to prevent them from killing themselves.

Wicked animals always seek their own kind, and often band together for evil purposes. Figuier tells of three beavers that built for themselves a nice little home near a stream, and they had as a neighbour a respectable hermit beaver. The three called on their neighbour one day, and he received them cordially, and hastened to return their visit, when they pounced upon him and slew him, like human murderers, who had trapped their victim.

From all these we learn that Nature is filled with life-saving and life-furthering adaptations. Just as in the human drama we find deceit, disguise, mask, trickery, bunco and bluff, all forms of cheating and clever deceptions, so it is precisely the same in the animal world, though man is little informed on Nature's real ways.

XIV

AS THE ALLIES OF MAN

"Who, after this, will dare gainsay
That beasts have sense as well as they?
For me--could I the ruler be--
They should have just as much as we,
In youth, at least. In early years,
Who thinks, reflects, or even fears?
Or if we do--unmeaning elves--
'Tis scarcely known e'en to ourselves.
Thus by example clear and plain,
We for these poor creatures claim
Sure sense to think, reflect, and plan,
And in this action rival man:
Their guide--not instinct blind alone,
But reason, somewhat like our own!"

The wonderful world in which we live is full of animal life. In the great forests, under the ground, on the steep mountainsides, in the depths of the oceans, rivers, streams, from the frigid north to the torrid south, in the parched deserts, are animals of every size, colour, and form, all of which are, in their general form, adapted to their peculiar places in nature. Their lives and habits undeniably demonstrate proofs of divine wisdom, intelligence, and beneficence. In fact they show an aptitude in many arts and sciences second only to that shown in man.

The reason that animals are often held in such low esteem by the world of science, is because people are apt to look upon them as natural mechanisms and overlook what they are doing and feeling. The propounders of false statements which attribute every act of an intelligent animal--second only to man and his faithful ally--as due to instinct only, deal with metaphysical reasoning. They have never considered the innumerable and irrefutable facts of animal life which no acuteness of analysis and pure thinking can ever explain. Most of these narrow, bookish men deny to animals capabilities which every country schoolboy knows they possess. It is no exaggeration to say that animals exist which sing, dance, play, speak a language, build homes, go to school and learn, wage warfare, protect their homes and property, marry, make laws, build moral codes, in fact, do everything that is generally attributed to man.

In comparing man and animals scientists are prone to ascribe to man as a whole the faculties which only the best trained and most talented possess. They fail to consider our cannibal brethren, such as are found among the Dyaks on the Island of Borneo, whose chief articles of adornment in the house are heads of murdered men, and whose savage and fiendish ways would put to shame a civilised animal. They forget how long man lived on this earth before he even learned to make fire by chipping flints.

Since the beginning of time animals have been the friends and allies of man. From the very earliest ages they have in innumerable ways been associated with historical events, and with the laws, customs, superstitions, and religions of all nations of the universe. Love, devotion, gratitude, the sense of duty, as well as all the lower passions of hatred, revenge, distrust and cunning are their heritage. Only an egotist who has known them in books only, and knows nothing of their mentality and brain power, would dare say that they are governed solely by instinct. Cases of animal suicide, following some deep disgrace among them, are not uncommon.

From the Bible we learn that God frequently employed animals as agents to dispense His providence. Bullocks, sheep, goats were used by the Jews in their religious services, while a disobedient prophet was killed by a lion. Balaam was rebuked for his cruelty by an ass; and David even called upon the animals to aid in praising Jehovah! That we may learn real gratitude for common mercies Isaiah says: "The ox knoweth his owner, and the ass his master's crib," *etc.* When the city of Nineveh was threatened, God had pity on it, because there were many cattle there. The Saviour compared his own earthly condition with that of certain animals: "The foxes have holes," *etc.* He called himself the 'Good Shepherd,' and his followers were sheep who knew his voice. John the Baptist referred to Him as the 'Lamb of God'; while John, the beloved disciple, when on the Isle of Patmos, saw the "throne of God in heaven, and before it a lion, a calf, a man, and a flying eagle."

The first beginnings of co-operation between men and animals must have begun by the approach of certain less timid animals, which felt that better conditions for them and more food could be obtained near human habitations, and perhaps, more protection from dangerous animals. Or it may have begun through the stupidity of certain animals who failed to realize the danger of man's proximity.

It seems that the secret ambition of all animals is to become the allies of man. This is demonstrated by the fact that most of them have gone near the villages and towns, and, consequently, there are comparatively few remaining in the heart of the big forests. Under the true state of conditions man should live in harmony with these animal brothers, with mutual trust and respect existing between them. That would mean, of course, that man would have to show a little more kindness to them. For while he is their true sovereign, he abuses the privileges of his sovereignty in untold ways, and up to the present time only a few animals, like the dog and horse, have been fully recognized as his allies.

All the others, with few exceptions, have shown a desire to become more closely united with man, and yet during the thousands of years of man's rulership over the beasts, he has been able to makc allics of only about sixty. This regrettable fact speaks for itself--showing that man has long abused his trust.

Warfare, as it is waged to-day, demonstrates that notwithstanding man's vast number of scientific aids, animals are still invaluable. The innumerable mechanical and electrical devices unknown ten years ago, such as enormous rapid-firing guns, walking "Willies," wireless machines, traction engines, smokeless and noiseless powder, silent-sleepers and tear-bombs, all of these have greatly increased man's power of offence and defence, yet with all these ultra-modern improvements, animals are absolutely essential in waging a successful war.

In military circles there is an ever-increasing demand for well-trained army horses, sound in mind and body and educated in modern campaigning. Above all, an army horse must be dependable, must love his soldier-master and must know absolute obedience to orders. Every army horse has to pass an examination and prove his worth before he is enlisted into the service.

The largest of the mountain guns used in Italy against the Austrians were drawn up the steep mountains by mules. Another 75-millimetre gun for mountain warfare is taken to pieces, into four parts, and each piece is separately packed on a mule.

The United States cavalry has the best trained war horses in the world; many of them actually understand the complicated commands of their masters. These horse soldiers have the insignia, U. S., branded on the hoof of the left forefoot, and the other animals in camp, on the shoulder.

When a horse arrives at a regiment he is assigned to a troop according to colour, size, weight and mental efficiency, and later he is permanently assigned to a man. Under no conditions is he interchanged or even ridden by another than his master, and it is astonishing the tremendous affection that oft-times springs up between the two; in many instances horses have been known to seek out their masters among hundreds of soldiers.

On the European battlefields, near which there are few or no railroads, animals have been the principal means of transportation, elephants, camels, horses, mules and oxen being chiefly used for this purpose. The Italian armies have used numerous teams of mountain-trained bullocks to draw loads up the mountains, and, while they cannot ascend roads as steep as those which the mules climb, they are very valuable for heavy loads. These bullocks work faster than an army mule, for a mule will never hurry. As the old darkey once said, "De mule warn't born fer to hurry; not even a torpedo would make him move one step farster!"

Elephants have been used to a small degree in the armies of Europe. While they are splendid workmen, they are dangerously subject to stampede, and one stampeding elephant can do much harm in an army.

The British army has used quite a few trained elephants from India in their ranks. They are especially employed to rout the enemy from small forests. Breaking through bushes, crushing underbrush, and pulling up small trees is their specialty. They make splendid bulwarks for soldiers, and when an army is marching through a forest, are invaluable in clearing the way. A British officer declared that one trained elephant is more valuable than a half-dozen traction engines.

Far the most interesting and curious use to which an animal is subjected is the use of camels chosen and trained because of their strange colouring and height. Small groups of them have been stationed among clumps of acacia trees with a spy mounted on the animal's neck. This is the safest place a person could be, for the camel or, in like manner, the giraffe, standing with only his head above the small trees, looks precisely like a bit of the foliage in the distance.

Camels are especially good for desert warfare, because they can go without water so long and can easily carry loads weighing from 400 to 500 pounds. In the last Afghan campaign the British lost over 50,000 camels and in the Great War they have had more than 60,000 in army service in Egypt. Camels are especially used for transportation purposes. The British capture of Jerusalem was greatly aided by these desert allies. Large numbers of oxen have been used in the French army. They do not balk at autos and know no fear of shells.

One of the greatest allies of the animal kingdom in warfare is the dog. These allies are trained to aid relief parties on the battlefields, and many of the ambulance men have their splendidly trained dogs for seeking out wounded soldiers among the dead. They are also trained as guards and watch-dogs and they become marvellously clever when used near the firing lines. They carry water in the trenches and are trained in packs to dismount enemy motorcyclists by pulling them from their machines. Dogs also make splendid scouts, and excellent and reliable messengers when not required to go too far.

These faithful friends of man, according to Buffon, are far more easily taught than man, and more easily led "than any of the other animals, for not only does the dog become educated in a short time, but even adapts himself to the habits of those who control him." According to circumstances, a dog may become a soldier, messenger, water-carrier, or guard.

THE ESQUIMO-DOG IS MAN'S GREATEST FRIEND IN THE FAR NORTH.

American Museum of Natural History, New York

CHIPMUNKS ARE AMONG THE MOST EASILY TAMED OF MAN'S WILD FRIENDS, AND THEY EVEN SEEM FOND OF HUMAN COMPANIONSHIP.

Not the least among the uses of war dogs is the curious practice of sending them into the enemies' lines of cavalry to convey fire in order to terrorise the horses and throw them into confusion. This practice has been quite common in the past. Each dog is dressed in a cuirass of leather and on his back is carefully strapped a pot of boiling, blazing tar. Nothing so terrorises horses as the sight of approaching fire.

A small but valuable ally to man is the ferret. This little creature has come into prominence more particularly during recent years, when the rat infested trenches have made his services invaluable. These Hun-like rats, devouring and devastating in their thirst for human blood, would have forced the abandonment of many a front line trench but for the aid of these trained ferrets, thousands of which have been daily employed on the battle fronts.

The immense services rendered by carrier pigeons in the battle of the Marne, not only to the military authorities, but also to the public at large, will cause the civilised world to pay more attention to the importance of these birds in the future. They carried all kinds of messages to and from Paris during this memorable battle; in fact, they have been used in all the battles as invaluable messengers.

Small animals, such as mice, canary birds, guinea pigs and rabbits are used in trench warfare, because they are more sensitive than man to poisonous gases. It sometimes happens that hundreds of men must be rescued from a trench by three or four men. Each rescuer carries with him a canary bird in a small cage attached to his shoulder. And as long as these birds show no signs of distress the men are safe from gas poison. The birds soon become attached to their masters and seem to like the adventure of the trenches.

As time goes on, it is to be hoped that we will understand our animal brothers better, and that our old attitude toward the so-called "brutes" will be entirely changed. Heretofore we have greatly abused the zebra, for example, because of his wild disposition, ferocious humour, distrust of all power except that in his own legs, and his pronounced aversion to work.

Why should we reproach him for his wildwood philosophy? It is perfectly natural that any animal of his experience with man, and with sufficient brains, would have only contempt for all mankind. His native home is in Africa, and his human associates, if they are human, have been the Hottentots, the Namaquois or the Amazoulons--the most impossible and hideous people on the earth. Since his babyhood days he has seen nothing but cannibalism and carnage among the savages; and since his transportation to Europe by a strange occurrence of horrible circumstances, he has been the subject for all kinds of barbarous punishments which man has seen well to heap upon him. The zebra is not of the mental calibre to be suddenly seized with love for the human species and its civilisations! And the human species is astounded and thinks the zebra stupid and wicked. He may be both, but his wisdom is undeniable when it comes to trusting humanity, and his wickedness is small in comparison to man's terrible cruelties. He should be awarded a medal for wisdom! For man is far the greater ass of the two!

He roams the wild prairies where the fields need no ploughing. There he finds an abundance of grass and fresh water along the streams. No loud cursing and swearing ever greets his ears, nothing but the sweet song of the wild birds. And his children romp and play with him, free as the winds that blow. Of course, he has enemies even there, and so he uses camouflage by painting himself in attractive stripes, so no one can see him at a distance. Even Solomon should have praised his wisdom!

In the beginning God created man, and not long after gave him as his policeman, the dog. And the obedience, friendship and devotion of the dog to his master has been unending. The dog discusses no questions of right or wrong, his only duty is to obey. This he does without a murmur. He is the greatest testimony to man's civilisation, the first and the greatest element of human progress. Through his co-operation man was elevated from the savage to the state of the civilised. He made the herd possible. Without him there could have been no herd, no assured subsistence of food and clothing, no time to study and improve the mind, no astronomical observations, no science, no arts, no automobiles, no airships, no wireless telegraphy--nothing. The East is the home of civilisation, because the East is the home of the dog.

A young hound knows more about tracking game or scenting the enemy after six months' practice than the most skilled savage after fifty years of study. The dog has so aided mankind as to give him more time for study and self-improvement. Thus began the arts and sciences. An interesting, and we believe original observation, of the influence of the dog on peoples is that wherever the dog is found, especially among the shepherd peoples, such as the Chaldeans, Egyptians, Arabs, Tartars, and Mongols, cannibalism is unknown. This is due to the fact that the dog enables them to maintain the herds which supply them with milk, food, and clothing, thus preserving them from the criminal temptation of hunger.

The Indians of North America never refrained from roasting their enemies until they made allies of the horse and dog. Humboldt proves the lively regret held by one of the last surviving chief lieutenants of the war-like Tecumseh whom he asked about a certain American officer who took part in the fight. "Uh!" replied the Indian, "I eat some of him." "Do you still eat your enemies?" asked Humboldt. "No," replied the Indian. "Big dog catch heap meat for me!"

Surely no animal could be more uncivilised or cannibalistic in its desires than man! Spinoza believed, however, that benevolence in animals consisted only in their kindliness and friendly feeling for each other and that we should expect nothing more of them. A good cow, so he thought, was one that was kind to her calf, however ferocious she might be toward human children. But we do not accept this standard of goodness, nor believe that animals' kindness extends only to their own tribes. Their lowest standard of life is no worse than the cannibalism existing among the lower tribes of uncivilised man, which is one of the highest ideals of tribal life. The greatest hero among our savages is the one that can put the most enemies to death.

Many animals seem to have a social instinct and a moral sentiment toward man. They try to break the old bonds of distrust between their master and themselves. This is especially true of the puma, second to the largest of the big cats of the Americas, which seems to love the society of man, and seeks not only to be near him, but to protect him from the attacks of the much-dreaded jaguar. A civil engineer tells the story of an experience he had while journeying up one of the big South American rivers by boat. At their nightly encampments one of the passengers on board was an old miner who insisted on sleeping in a hammock suspended between two small trees. His weight was sufficient to bring the hammock almost to the ground at its lowest curve. One morning, his friends inquired how he had slept, and he complained that "the frogs and small animals had made so much noise under the hammock that he could not sleep." One of the Indian servants roared with laughter, as he said, "Uh, 'tiger' sleep with old man last night. He watch him!"--tiger being the Indian term for the puma. Careful searching revealed the footprints of an immense puma, and that he had evidently lain directly under the hammock. The noise which had kept the old man from sleeping was the purring of the animal, pleased over the privilege of sleeping so near a man. These Guiana Indians know the ways of the forests, and have a special liking for wild animals. This entire absence of fear in the puma is the same as exhibited by the tame house cat.

Many animals seem fond of human companionship, and are easily tamed. My sister raised a small red deer in Texas, and he became so perfectly tame that he would follow her wherever she went, and would even take food from her hand. In Yellowstone Park the deer are so tame they will come into the yards to get food, while the brown bears approach the hotels like tramps, and many of the smaller animals are perfectly fearless. At the Bronx Zoological Gardens, and the London Zoo, the animals have lost all fear. They seem to realise that they have no power to escape and depend entirely upon man for their daily food. But, of course, their conditions are artificial, hence such conclusions as we may draw as to their normal attitude toward man do not necessarily indicate the innate character of their wild kinsmen. We occasionally find, for instance, that in unsettled regions like parts of Mexico and South America, where animals are plentiful and man's influence largely absent, they are found to be particularly ferocious, yet even then lions and leopards rarely attack men unless disturbed in some unusual way.

Quite a few naturalists and scientists believe that the animals' love for man was acquired and not natural. But if this be true, how did the very early tribes of men escape destruction at the hands of the wild beasts which were far more numerous than at present? The animal kingdom was evidently impressed by the power of man at a very early stage of its development, but in just what manner or what period of time this came to pass is not known.

If we regard the conflict as merely between two great groups of animals, surely the animals should have won, and man would have disappeared from the face of the earth. The fact that he did not, and that he became master of the animals, is presumptive evidence that man exceeded the animals in intelligence.

Primitive man could have lived in no other way than by "his wits." For he was not nearly so well equipped for defence as are the monkeys of to-day. Their greatest power is in the ability to use their arms and hands in swinging rapidly from branch to branch. This gives them an advantage over all tree-climbing cats. They are very proficient in throwing stones and other missiles. This is dumbfounding to other animals. Of course, their intelligent and quick-witted methods of defence, menace, guard-duty, and loyalty to tribe makes them great warriors, and enables them to survive even the onslaughts of their greatest enemy and nightmare of every non-carnivorous animal--the harpy eagle!

Through the necessary adjustments growing out of the close relationships of men to animals, the mental faculties of both have been greatly stimulated and advanced. The least developed races seem to be in such places as Tierra del Fuego, where there are no savage animals, and, therefore, no inducement for man to arm and defend himself. The Pygmies of Central Africa are mighty hunters, otherwise they could not survive. Even the Esquimaux are masters of the great polar bears and other northern animals.

In the wilds of Africa, where animals have had a terrible struggle for existence, not only against disagreeable climatic conditions, but all kinds of fellow-foes as well, we find the nkengos have attained a civilisation that almost equals that of our savage brothers. And these pale-faced little beings, with their wrinkled, care-worn, parchment-like skins, remind one of ill-treated, white, human-dwarfs. Their name, nkengo, means wild animal-men, and when tamed they actually make excellent family servants for men.

These closest allies of man live in tall bamboo trees, and are so curiously human that when seen walking around hunting berries, nuts, and fruits, talking in guttural, chattering tones, like old fisher-women, no one could doubt even their kinship to man.

Their children assemble in groups to romp and play under the guardianship of either one of their mothers or grandmothers; while the men forage for food, and watch for enemies. It is not uncommon to see an aged, half-decrepit nkengo lying on a bed of sticks in a tall tree. Here he eats only green leaves and bits of fruit brought him by some kind friend, being far too weak to hunt for food himself, and furthermore, fearing an attack from his mortal enemy, the leopard.

If the colony decides to move to other territory, either because of enemies or the scarcity of food, they all assemble and hold a farewell gathering in which there is much mourning and apparent grief at forever leaving their aged kin to the fate of the wilds. If they are possibly able to walk, they are given patient assistance in travelling along. Sometimes, when they are deserted, sympathetic friends return for days with berries and koola nuts, until at last the colony has gone so far away that none dare return alone, in which event these helpless superannuated members are left to die in their lone tree-top beds.

Many of these beds are as well made as the tree-beds of human beings, and even better than the beds of the savage Dyaks of Borneo. They are usually located in tall trees, inaccessible to leopards and out of reach of their most dreaded of all enemies, the terrible hordes of war-ants. From these nothing escapes--not even elephants and tigers.

The arrival of a baby to these nkengos is of far more importance in their tree-top village, than in a human city. Each of the female relatives, and also the aged males, takes special interest in the new-comer, and they chatter around his little grape-vine cradle with much enthusiasm, shaking their heads and delicately handling his tiny hands and toes as though he were the baby of a king.

This baby is much stronger and quicker to learn than human babies; for when he is only two days old he is able to cling to his mother, so that she can carry him with her on her hunting trips. If he becomes too noisy from sheer delight when she is travelling through the forest with him, she slaps him, in an attempt to quiet him, lest the leopards get him.

At night he sleeps snugly by his mother's side in the great tree-bed, and she never allows him to crawl out of her arms for fear that he fall to the depths below. She loves him dearly, and watches with human eagerness for his first tooth. He loves his mother and will stand for hours while she dresses his hair; or lie on her breast as she rubs his little back.

These wild-children are always ill-tempered and self-willed. No human mother has to show more patience and love than does the nkengo mother. She takes the greatest delight in his first efforts at climbing and hunting, and for hours she and his admiring relatives will watch him attempting to climb a cocoanut tree. Sometimes she will climb just behind him to catch him if he falls or becomes frightened.

His arms soon become very powerful, for he is constantly swinging, climbing, and exercising by hanging from a bough with one hand while he pulls himself up with the great power of his muscles. He is able to gather koola nuts long before his jaws are strong enough to crack them; so his fond mother cracks them for him until his hands and mouth are stronger. Like all babies, his ambition is to be big and strong like his father.

Some of the apes are most intelligent and human, and, as allies to man, are more desirable than certain of the human savages. Dr. Livingstone, in his *Last Journals*, describes one he first discovered. "Their teeth," he says, "are slightly human, but their canines show the beast by their large development. The hands, or rather the fingers, are like those of the natives. They live in communities consisting of about a dozen individuals, and are strictly monogamous in their conjugal relations, and vegetarian, or rather frugivorous, in their diet, their favourite food being bananas." The natives where these apes live are cannibals, and Dr. Livingstone says, "they are the lowest of the low." One of their number, who had committed a great murder, offered his grandmother "to be killed in expiation of his offence, and this vicarious punishment was accepted as satisfactory."

Thus it is evident that certain of these wild-creatures--like the sokos--have a more correct conception of justice than their human associates, the savages. At least the animals do not make the innocent suffer for the guilty, and give their lives unjustly. Should a soko try to take another's wife he is publicly punished by the tribe. These animals have a great sense of humour and fully enjoy a practical joke. Strangely enough, they never attack women and children, but if any man approaches them with a spear or gun, they try to rush upon him, often at the expense of their own life, and wrest the weapon from him. Most of them are exceedingly kind and civilised in their actions, and natives always say, "Soko is a man, and nothing bad in him."

Often they kidnap babies and carry them up into trees. But these are never harmed and the apes are ever ready to exchange them for bananas. The robbery is, no doubt, for the purpose of extortion. If perchance one of their children is stolen, the entire forest sets up a scream and wail until it is returned. Old hunters and travellers say that they would rather steal the child of a native savage than to take one of the sokos. If one of the soko children disappears, and they do not know what became of it, they immediately send out detectives throughout the country to seek for it. And woe be the home where a stolen soko baby is found!

But man has one great power--a far more potent ally than he has in his animal friends--the use of fire. Unquestionably to the minds of animals it is a supernatural power. They cannot create it, understand it, and it is very doubtful if they can yet use it to advantage. How marvellous is this thing--fire! That great blazing pillar of cloud that destroys all, and leaves nothing to show where it has taken its enemies! To animals it springs up wherever man rests his head, and protects him while he sleeps. It is always with him, and its presence for untold ages has brought terror to all of them.

Not a few reports tell us that certain of our animal allies among the monkeyfolk of South Africa use fire. This may not be true; but it is probable that the time is near at hand when the wild baboon-men of the woods will learn to make and use fire just as we have done.

Enough instances could be shown illustrating animals as man's allies to fill an entire book, but a sufficient number have been adduced to show how truly they are our allies, helpers, and protectors just as we are theirs, only their mode of manifesting it is different. We have shown the absolute fallacy of the old belief that animals lack mentality, and that all their acts of kindness are based upon self-love and personal gain, and have seen that in proportion to their opportunities in life, they have quite as much mentality and brotherly love for each other and mankind as is found among our lower savages. We have seen that among animals as among men, individuals will give their lives for their fellows, serve the weak and timid, and demonstrate the highest and holiest feelings of which true souls can be capable, and always share equally with man the burdens that fall upon themselves and their human allies. And the time is already here when man should protect his animal friends more, and teach them through human kindness not to fear him. But this can only be done when he is willing to treat them as fellow beings only a little below him in the scale of existence.

CHAPTER XV

THE FUTURE LIFE OF ANIMALS

"Ah, poor companion! when thou followedst last
Thy master's parting footsteps to the gate
Which closed forever on him, thou didst lose
Thy best friend, and none was left to plead
For the old age of brute fidelity.
But fare thee well. Mine is no narrowed creed;
And He who gave thee being did not frame
The mystery of Life to be the sport
Of merciless man. There is another world
For all that live and move--a better one!
Where the proud bipeds, who would fain confine
Of their own charity, may envy thee."

--SOUTHEY (on the death of his dog).

The old belief is still prevalent that the Bible teaches that of all living creatures man alone is immortal. This erroneous belief springs out of man's egotism, however, and is not substantiated by the Scriptures. Among many of the Old Testament writers we find that immortality was assured for neither man nor animals; whereas, with the larger revelation of the New Testament, immortality is no longer questioned for any living creature.

There are, of course, many supposedly intelligent people who deny to animals the power of reason, and attribute all their marvellous powers and abilities to blind instinct. It is, therefore, not the least bit surprising that the vast majority of people believe that when an animal dies, its life principle dies also. The animating power, they believe, is destroyed, and the body returns to the dust.

These mistaken conclusions are largely, if not wholly, due to two passages of Scripture, one of which is in the Psalms and the other in Ecclesiastes. The one most often quoted, from the Psalms, runs in the authorised version: "Nevertheless, man being in honor, abideth not; he is like the beasts that perish." This verse is frequently quoted as decisive of the whole question. The other passage, which is found in Ecclesiastes, reads: "Who knoweth the spirit of man that goeth upward, and the spirit of the beast that goeth downward to the earth?"

It is upon the authority of these two passages that we are supposed to believe that when an animal dies, its life has gone forever, departed, expired. In this new age of thought and discovery, we do not attempt to explain a passage of Scripture, no matter how simple it may appear to be, without referring to the original text, that we may see if the translator has kept the true sense of the words and adequately expressed their significance, remembering that words often change their meaning, and that the original use of a word may have conveyed exactly the opposite meaning to that which we at present attach to it.

But if we accept the passage just as it stands, with the literal meaning of the words as is usually understood, there is but one conclusion--animals have no future life. Death ends all for them. But, on the other hand, if we are to take the literal interpretation of the Bible only, we are forced to believe that man, as well as the animals, has no life after death. Surely the book of Psalms is full of examples to support this literal interpretation. For example, "In death there is no remembrance of thee: in the grave, who shall give thee thanks?" Again, "The dead praise not the Lord, neither any that go down into silence." Or, "His breath goeth forth, he returneth to his earth; in that very day his thoughts perish." These quotations could be greatly added to, and if taken in their literal sense, we would reach but one conclusion--death ends all for every living creature! Nothing in all the literature of the earth could be more gloomy and discouraging than these quotations with numerous others that contemplate death. Yet, vain man takes one little passage that seemingly denies a future life to animals from the same book that many times over denies a future life to mankind; in fact, there are five times as many Scripture passages claiming for man that all ends in death as there are for animals. Over and over we are told that those who have died have no remembrance of God, and cannot praise Him. The Bible speaks of death as the "land of forgetfulness,"--the place of darkness, where all man's thoughts perish. Nothing more than this could be said of the "animals that perish!"

Other Biblical writers referred to mankind as those who "dwell in houses of clay," and Job says: "They are destroyed from morning to evening; they perish forever, without any regarding it." In another place he says: "As the cloud is consumed and vanisheth away, so he that goeth down to the grave shall come up no more." Again he speaks of "the land of darkness and the shadow of death," and says: "Man dieth, and wasteth away: yea, man giveth up the ghost, and where is he? As the waters fail from the sea, and the flood decayeth and drieth up: so man lieth down, and riseth not." Job laments the pitiable conditions of his life, and complains that life was ever granted to him, and that even death can bring nothing to him except extinction.

Yet, if we examine Ecclesiastes, the book in which we find the single passage upon which many people base a belief in the non-future existence of animals, there are passages which are really no more positive as to the future of mankind. For example, "I said in my heart concerning the estate of the sons of men, that God might manifest them, and that they might see that they themselves are beasts. For that which befalleth the sons of men befalleth beasts; even one thing befalleth them. As the one dieth, so dieth the other; yea, they have all one breath, so that a man has no pre-eminence over a beast: for all is vanity. All go unto one place; all are of the dust, and all turn to the dust again." Again it is said: "For the living know that they shall die, but the dead know not anything, neither have they any more a reward, for the memory of them is forgotten;" and "Whatsoever thy hand findeth to do, do it with thy might; for there is no work, nor device, nor knowledge, nor wisdom in the grave whither thou goest."

By interpreting these words literally, there is but one conclusion relative to a future spiritual life, namely, that there is absolutely no distinction between man and his "lower brother" animals, and that when they die they all go to the same place. It is emphatically said that after death man knows nothing, receives no reward, and can do no work. Job has the same gloomy strain running through his writings, and Ecclesiastes gives a most morbid and gloomy view of death.

However, no modern Biblical scholar accepts these passages in this literal light, for it is known that they were written symbolically, or as parables, and were not intended to be literally interpreted. They have a spiritual significance. We are, however, not interested here so much with this spiritual sense as we are with the literal implication of the translation. Therefore, according to this literal meaning of the two texts, if we accept them to prove that animals have no future life, we are forced to believe by at least fourteen passages, of equal if not greater power, that man shares their same fate after death. No man has a right to select certain passages from the same book of the Bible and say that they shall be accepted literally, and that other passages of equal merit shall be interpreted otherwise. They must all be treated the same.

All scholars are familiar with that remarkable eleventh book of Homer's Odyssey, known as the Necromanteia, or Invocation of the Dead, and in it Ulysses descends into the regions of the departed spirits to invoke them and obtain advice as to his future adventures. One commentator says: "He sails to the boundaries of the ocean, and lands in the country of the Cimmerians, who dwell in perpetual cloud and darkness, and in whose country are the gates leading to the regions of the dead." All is darkness, discontent, hunger; nothing is said of virtue, wisdom, beauty, happiness. Only bitter gloom! No wonder this heathen poet considered, with such views of a future life, sensual pleasures as the chief object of this life.

The following dialogue between the inhabitants of the earth and the dweller in the regions of the dead--between Ulysses and Achilles--is remarkable for its horrible depiction of the future life:

"Through the thick gloom his friend Achilles knew,
As he speaks the tears dissolve in dew.
'Comest thou alive to view the Stygian bounds,
Where the wan spectres walk eternal rounds;
Nor fear'st the dark and dismal waste to tread,
Thronged with pale ghosts familiar with the dead?'
To whom with sighs, 'I pass these dreadful gates
To seek the Theban, and consult the Fates;
For still distressed I roam from coast to coast,
Lost to my friends and to my country lost.
But sure the eye of Time beholds no name
So blessed as thine in all the rolls of fame;
Alive we hailed thee with our guardian gods,
And, dead thou rulest a king in these abodes.'
'Talk not of ruling in this dolorous gloom,
Nor think vain words (he cried) can ease my doom.
Rather I'd choose laboriously to bear
A weight of woes and breathe the vital air,
A slave for some poor hind that toils for bread,
Than reign the sceptered monarch of the dead.'"

Yet, even this outpouring of hopeless words by the heathen poet is encouraging when compared to the writings of the Psalmist, of Solomon or Job, for those who have gone beyond the grave still have memory, an interest in their friends on earth, love and desire. But no such hope exists for man, if we are to accept literally all the passages of Scripture which have been quoted. By such interpretation, man passes after death into eternal darkness, forgetfulness, silence, "where there is no work, nor device, nor knowledge, nor wisdom--where even his very thoughts perish." If these particular passages are to be accepted as final on the subject, there is no future life for either man or animal. They are too definite to admit of any interpretation that might soften or alter their meaning.

It may be shocking to some to compare the belief of an ancient Greek and the teachings of a Latin Epicurean with the sacred writings of the Bible. Yet, it may be even more startling to point out that some of the teachings of the Epicurean sensualist are quite as good as some of those of the writers of the sacred texts, and that those of the Greek poet are far better and more spiritual! There is no denying that these are the facts, if we are to be bound by literal interpretation, unless we throw to the winds all reason and common-sense.

This leads us back to the point previously mentioned; and we must determine if the authorised version gives a full and truthful interpretation of the Hebrew original. Even a man who does not pretend to scholarship knows that it does not. The word "perish," for example, is not found at all in the Hebrew text, nor is the idea expressed; the words which our translation twice renders as "beasts that perish," is, in the original Hebrew, "dumb beasts." By comparing a number of the translations of the Psalms, into various languages--Psalm XLIX, for example--we find that few, if any, of them suggest the idea of "perishing" in the sense of annihilation. First, let us consider the Jewish Bible, which is acknowledged to be the most accurate translation in the English language, and carefully read it. In verses 12 and 20 of the above Psalm, where the passage is found, the translation reads: "Man that is in honour, and understandeth this not, is like the beasts that are irrational." In a footnote the word "dumb" is offered as an alternative for "irrational." Brunton's translation of the Septuagint is similar, and reads: "Man that is in honour understands not, he is compared to the senseless cattle, and is like them." Wycliffe's Bible, which is translated from the Vulgate, reads thus: "A man whanne he was in honour understood it not; he is compared to unwise beestis, and is maad lijk to tho." The "Douay" Bible, put forth by the English Catholic College of Douay and which is received by the Catholic Church in England, gives the passage: "Man, when he was in honour, did not understand; he hath been compared to senseless beasts, and made like to them." Many other versions might be cited, and very few of them even suggest the idea of annihilation. If, for argument's sake, we suppose that the word "perish" has been correctly translated, it by no means follows that annihilation is signified. Read, for example, the tenth verse of the same Psalm in our authorised translation: "For he seeth that wise men die, and likewise the fool and the brutish person perish, and leave their wealth to others." Certainly no intelligent person would interpret this passage as declaring that the wise and the foolish and the brutish have no life after the body dies.

It is plain, therefore, that we may dismiss forever the idea that the Psalmist believed the beasts had no future life, and the citation may be rejected as absolutely irrelevant to the subject, and the only one that appears to make any definite statements as to the future life of the lower animals. Every student of the Bible will at once recognise how necessary it is that the original meaning of the Hebrew text should be known, and that the Psalmist should not be accused of setting forth a doctrine of such great importance, whether true or false, when he may never even have thought or suggested it.

MEN CRUELLY TAKE THE LIVES OF THESE DENIZENS OF THE WILDWOOD, REJOICING IN THEIR SLAUGHTER, BUT THE ANIMAL SOUL THEY CANNOT KILL.

TWO PALS. THERE IS BETWEEN MAN AND DOG A KINSHIP OF SPIRIT THAT CANNOT BE DENIED.

Having disposed of the possibility of a misunderstanding of the real meaning of the "beasts that perish," let us consider the quotation from

Ecclesiastes, the only one that refers to the future state of animals. "Who knoweth the spirit of man that goeth upward, and the spirit of the beast that goeth downward to the earth?" We find an admission here that, whether the spirit ascends or descends, man and beasts alike have the immortal spark. The Hebrew version is precisely the same as our authorised translation. Read, not an isolated verse, but the entire passage:

"I said in mine heart concerning the estate of the sons of man, that God might manifest them, and that they might see that they themselves are beasts.

"For that which befalleth the sons of men befalleth beasts; even the one thing befalleth them; as the one dieth, so dieth the other; yea, they have all one breath; so that a man hath no pre-eminence above a beast: for all is vanity.

"All go to one place; all are of the same dust, and all turn to dust again.

"Who knoweth the spirit of man that goeth upward, and the spirit of the beast that goeth downward to the earth?

"Wherefore I perceive that there is nothing better than that a man should rejoice in his own works; for that is his portion; for who shall bring him to see what shall be after him?"

These verses tell their own story. It matters little whether Solomon wrote this book in his later years; it is, in any event, the confession of one who has had all the good things of this world, and who saw the emptiness of them all, and who sums up life with the words "Vanity of vanities, all is vanity." Finally the author ironically advises his readers to trust only in the good of their labour.

Thus it is shown that the quotation from the Psalms in no way justifies the belief in the annihilation of beasts, and that the one from Ecclesiastes has been entirely and wrongfully misunderstood and interpreted. In no way do the Scriptures deny future life to the lower animals, but in all ways, if intelligently understood, imply that man and beasts have, equally, a share in a future life beyond the grave.

As we have found out that the Scriptures, contrary to the popular belief, do not deny a future life to our lower brethren, the animals, let us see if they actually declare a future world for them in the same way that they do for man. Man's immortality, as we know, is taught in the Old Testament rather by inference than by direct affirmation. This is possibly due to the fact that the writers of the manifold books, which were at a late date selected from a large number and made into one big volume which forms our Bible, thought as a matter of course that man lived on after death, and never thought it necessary to assert that which every one knew.

But if we accept the teachings of the Old Testament, inference gives much stronger testimony to the immortality of animals than it does to the immortality of man, for while in neither case is there a direct assertion of a future life, yet there is no direct denial of future life to the animals, as has been shown to be the case with man.

All Divine Law includes a protection for the beasts, and the laws of the Sabbath were in essence a spiritual and not only a physical ordinance. The ancient Scriptures have innumerable provisions against mistreating or giving unnecessary pain to the lower animals; and these provisions stand side by side in the Divine Law with those which speak of man. Note, for example, the prohibition of "seething a kid in its mother's milk." Again, there is a statement that the ox in treading out the corn is not to be muzzled, lest he suffer hunger in the presence of food which he may not eat.

In the following sentences from the Book of Jonah, it is plainly seen that the Deity has not failed to take notice of the animals: "And should I not spare Nineveh, that great city, wherein are more than six score thousand persons that cannot discern between their right hand and their left hand; and also much cattle?" Again, in the Psalms, "Every beast of the forest is mine, and the cattle upon a thousand hills. I know all the fowls of the mountains: and the wild beasts of the field are mine." Other passages that proclaim God as the protector of beasts, as well as man, might be cited, for the Bible makes frequent mention of them. Each of these Scriptures unquestionably proves that God has an interest in all His creatures, and that each shares His universal love.

No one can deny that Genesis, ninth chapter and fifth verse, refers to a future life for beasts as well as man; it is a part of the law which was given to Noah and which was the forerunner of the fuller law handed down through Moses: "Surely, your blood of your lives will I require; at the hand of every beast will I require it, and at the hand of every man; at the hand of every man's brother will I require the life of man." According to the Mosaic law, an ox which kills a man is subject to death, exactly as a human murderer. Why should the animal be punished by death, if he has no soul to be forfeited?

It should be remembered that while there are no Scriptural passages that definitely promise immortality to animals, there are many which infer it. Moreover, we should not expect to gain definite information on the subject from the Bible, for it was written for human beings and not for animals. If there are few direct references to the future life of man, surely there must be still fewer to that of animals!

But just as man has for countless ages had within himself an everlasting witness to his own immortality, so do we find that all who have really become acquainted with the lower animals, with their unselfishness, parental love, devotion to duty, generosity, wonderful mentality, and self-sacrifice--all those who know them realise that they are subject to the same moral law as man and share with him a future life.

Lamartine beautifully expresses a future hope for his faithful dog:

"I cannot, will not, deem thee a deceiving,
Illusive mockery of human feeling,
A body organized, by fond caress
Warmed into seeming tenderness;
A mere automaton, on which our love
Plays, as on puppets, when their wires we move.
No! when that feeling quits thy glazing eye,
'Twill live in some blest world beyond the sky."

Who can say that from the depths of the wide ocean, from regions unknown, and lands unexplored by man; from the remotest islands of the sea, and even from the far icy North, there are not animal voices ever rising in praise of our common Creator? The Bible says: "The Lord is good to all, and His tender mercies are over all His works," and, "All Thy works shall praise thee, O Lord,"--surely these endorse the above statements. And why should man define the limit of God's goodness, His love, care, and attention to the wants and needs of all His creatures?

The distinguished animal authority, Dr. Abercrombie, admitted that animals have an "immaterial principle" in them, which is distinct from matter. But he does not say that this principle, or soul, will live after death, as it is supposed to in man. However, many scholars both of ancient and modern times hold this opinion. Broderip, in his *Zoological Recreations* devotes much space in referring to ancient philosophers and poets, Christian Fathers, and Jewish Rabbis that have believed in the immortality of animals. The heroes of Virgil have horses to drive in the Elysian fields; the Greek poets gave to Orion dogs. Rabbi Manesseh, speaking of the resurrection, says, "brutes will then enjoy a much happier state of being than they experienced here," and a number of scholars, like Philo Judaeus, believe that ferocious beasts will in a future state lose their ferociousness. Among more recent scholars who hold this belief is Dr. John Brown, who boldly says: "I am one of those who believe that dogs have a next world; and why not?" The Rev. J. G. Wood said: "Much of the present heedlessness respecting animals is caused by the popular idea that they have no souls, and that when they die they entirely perish. Whence came that most preposterous idea? Surely not from the only source where we might expect to learn about souls--not from the Bible, for there we distinctly read of 'the spirit of the sons of man,' and immediately afterwards of 'the spirit of the beasts,' one aspiring, the other not so. And a necessary consequence of the spirit is a life after the death of the body. Let any one wait in a frequented thoroughfare for one short hour, and watch the sufferings of the poor brutes that pass by. Then, unless he denies the Divine Providence, he will see clearly that unless these poor creatures were compensated in a future life, there is no such quality as justice."

Eugene T. Zimmerman says: "I cannot help but think that my faithful dog, and playmate of my younger days, will have some form of a future life."

We do not recognise an absolute spiritual barrier of separation between man and animals. Man is an animal--the first of animals; but it does not of necessity follow that he will always continue to be so. By what right does he presume to deny a soul and a continued spiritual existence to lower animals? Are we not all of us fellows and co-workers, partakers of the same universal life, sharing alike a common source and destiny? This has always been the faith and insight of the child, whose simple wisdom we ever turn to for truth and guidance. And in our clearer realisation of the oneness of all life, we will extend to all creatures the Golden Rule, showing them the love and consideration we would have shown to us.

THE END